Feedback Control
for Computer Systems

Philipp K. Janert

O'REILLY®

Beijing · Cambridge · Farnham · Köln · Sebastopol · Tokyo

Feedback Control for Computer Systems

by Philipp K. Janert

Copyright © 2014 Philipp K. Janert. All rights reserved.

Printed in the United States of America.

Published by O'Reilly Media, Inc., 1005 Gravenstein Highway North, Sebastopol, CA 95472.

O'Reilly books may be purchased for educational, business, or sales promotional use. Online editions are also available for most titles (*http://my.safaribooksonline.com*). For more information, contact our corporate/institutional sales department: 800-998-9938 or *corporate@oreilly.com*.

Editors: Mike Loukides and Meghan Blanchette	**Indexer:** WordCo Indexing Services, Inc.
Production Editor: Christopher Hearse	**Cover Designer:** Randy Comer
Copyeditor: Matt Darnell	**Interior Designer:** David Futato
Proofreader: Julie Van Keuren	**Illustrators:** Philipp K. Janert and Rebecca Demarest

September 2013:   First Edition

Revision History for the First Edition:

2013-09-17:   First release

See *http://oreilly.com/catalog/errata.csp?isbn=9781449361693* for release details.

ISBN: 978-1-449-36169-3

[CK]

Table of Contents

Part III.   Case Studies

Part IV. Theory

Part V.   Appendices

Preface

This is a book about feedback control—not a topic that programmers (among others) tend to know much about. This is a pity, because feedback control was originally devised to solve a problem that should be all too familiar to software engineers, especially those who are working on enterprise systems. Feedback control is a way to make sure that large, complicated systems run reliably, even when subject to external disturbances, and to make efficient use of constrained resources.

If you are looking for a system that can spin up some additional servers when traffic in your data center spikes and take them down again when the rush is over, then you have come to the right place.

What Is Feedback?

Feedback works by constantly comparing the actual behavior of a system to its desired behavior. If the actual behavior differs from the desired one, a corrective action is applied to counteract the deviation and drive the system back to its target. This process is repeated constantly, as long as the system is running.

One appealing feature of feedback control is that it requires relatively little knowledge about the controlled system. As long as one knows which direction to "nudge" the system when it has gotten off course, one can build a feedback loop. For this reason, feedback is an attractive technique for controlling large, complex, and opaque systems.

Moreover, feedback systems are self-correcting even in the presence of external disturbances. Because the system's behavior is monitored and adjusted all the time, a feedback system naturally and automatically responds to changes in operating conditions. No need to make special provisions to activate additional servers for rush hour: the feedback controller will notice the increase in load and spin up further instances until the desired quality of service is met. It will also take the instances down again, once the rush has passed, in proportion to the lightening load.

One needs to be careful, though, since control actions that are either too large or improperly timed can "overcorrect" a disturbance. Instead of reducing the difference between the desired and the actual behavior, such control actions replace a deviation in one direction with another deviation in the opposite direction. In the worst case, the amplitude of these deviations grows with each step until the system becomes dysfunctional or, often literally, blows up!

To avoid this outcome, control theory has developed specific experiments for analyzing a system's behavior. The results from these experiments can then be used to design and tune control loops that are safe to operate and that will track a reference value accurately and reliably.

Why This Book?

Feedback control has a long and successful history in applications to electronic circuits, industrial processes, and vehicular engineering. But it can do more. Feedback is self-correcting, so it can keep a system on target even if conditions change unpredictably. Feedback requires only limited knowledge about the process to be controlled; hence it can be applied to situations involving complex and opaque systems, such as those typical of enterprise environments.

In this book we will study the application of feedback principles to several software engineering problems, such as the optimal sizing of a cache, the management of a server farm, the control of waiting queues or buffers, and more. In all these examples, feedback will help us to make efficient use of scarce or expensive resources and to do so in the face of changing conditions.

But applying feedback principles to computer systems raises different questions than one typically encounters in the traditional application areas. The laws describing the behavior of computer systems are much less constrained than those for systems in the physical world; so we will rely more on experimental measurements and phenomenological descriptions than on theoretical analysis. (This is similar to the situation one finds in the application of feedback methods to industrial processes.) At the same time, computer systems offer a greater variety of control signals than most physical assemblies do; we therefore have greater freedom in choosing the best signal to use and hence must be aware of the trade-offs involved. We will pay particular attention to such questions.

Feedback control has been barely explored as a design paradigm for software systems. I hope to convince you that feedback control has much to offer in this regard and is, in fact, the correct solution to many problems that software engineers commonly face.

How to Read This Book

It can be difficult for an outsider to learn about feedback. Textbooks and articles use specialized terminology and examples from existing application areas, which sometimes obscures the underlying concepts. The problems that arise in the classical application areas are not necessarily the same problems that are of greatest interest to programmers. One also needs to be aware that many textbooks on feedback control are primarily concerned with the mathematical manipulations that underlie control theory and so give less attention to conceptual development or practical implementation questions.

This book takes a different approach. Theoretical development, important and beautiful as it may be, has been relegated to Part IV in the back of the book. An introductory Part I introduces the concepts of feedback control, system dynamics, and controller design. Part II describes a variety of practical techniques for the implementation and tuning of controllers, and it also discusses some examples of "design patterns" for feedback loops. Part III consists of a collection of case studies: specific problems involving computer systems that are solved using feedback methods. For each case study, a number of different approaches and their trade-offs are discussed in some detail.

The case studies are the heart of the book. I suggest beginning with the introductory Part I in order to become familiar with the basic feedback concepts. Then tackle the case studies, diving into Part II (Practice) or Part IV (Theory) as the need for additional information arises. (The sequence of case studies is arranged roughly in order of increasing complexity.)

All case studies are realized as computer simulations, and the code is available (*http://examples.oreilly.com/9781449361693-files/*) from the book's website. The code is intentionally simple and straightforward so that it can be easily extended and modified. Experimenting with simulations is an excellent way to build intuition for the sometimes surprising behavior of closed-loop systems—and to build the necessary confidence that this feedback stuff really *works*!

Conventions Used in This Book

The following typographical conventions are used in this book:

Italic
> Indicates new terms, URLs, email addresses, filenames, and file extensions.

`Constant width`
> Used for program listings, as well as within paragraphs to refer to program elements such as variable or function names, databases, data types, environment variables, statements, and keywords.

`Constant width bold`
> Shows commands or other text that should be typed literally by the user.

`Constant width italic`
> Shows text that should be replaced with user-supplied values or by values determined by context.

 This icon signifies a tip, suggestion, or general note.

 This icon indicates a warning or caution.

Using Code Examples

This book is here to help you get your job done. In general, if this book includes code examples, you may use the code in your programs and documentation. You do not need to contact us for permission unless you're reproducing a significant portion of the code. For example, writing a program that uses several chunks of code from this book does not require permission. Selling or distributing a CD-ROM of examples from O'Reilly books does require permission. Answering a question by citing this book and quoting example code does not require permission. Incorporating a significant amount of example code from this book into your product's documentation does require permission.

We appreciate, but do not require, attribution. An attribution usually includes the title, author, publisher, and ISBN. For example: "*Feedback Control for Computer Systems* by Philipp K. Janert (O'Reilly). Copyright 2014 Philipp K. Janert, 978-1-449-36169-3."

If you feel your use of code examples falls outside fair use or the permission given above, feel free to contact us at *permissions@oreilly.com*.

Safari® Books Online

 Safari Books Online (*www.safaribooksonline.com*) is an on-demand digital library that delivers expert content in both book and video form from the world's leading authors in technology and business.

Technology professionals, software developers, web designers, and business and creative professionals use Safari Books Online as their primary resource for research, problem solving, learning, and certification training.

Safari Books Online offers a range of product mixes and pricing programs for organizations, government agencies, and individuals. Subscribers have access to thousands of books, training videos, and prepublication manuscripts in one fully searchable database from publishers like O'Reilly Media, Prentice Hall Professional, Addison-Wesley Professional, Microsoft Press, Sams, Que, Peachpit Press, Focal Press, Cisco Press, John Wiley & Sons, Syngress, Morgan Kaufmann, IBM Redbooks, Packt, Adobe Press, FT Press, Apress, Manning, New Riders, McGraw-Hill, Jones & Bartlett, Course Technology, and dozens more. For more information about Safari Books Online, please visit us online.

How to Contact Us

Please address comments and questions concerning this book to the publisher:

O'Reilly Media, Inc.
1005 Gravenstein Highway North
Sebastopol, CA 95472
800-998-9938 (in the United States or Canada)
707-829-0515 (international or local)
707-829-0104 (fax)

We have a web page for this book, where we list errata, examples, and any additional information. You can access this page at *http://bit.ly/feedback-control*.

To comment or ask technical questions about this book, send email to *bookquestions@oreilly.com*.

For more information about our books, courses, conferences, and news, see our website at *http://www.oreilly.com*.

Find us on Facebook: *http://facebook.com/oreilly*

Follow us on Twitter: *http://twitter.com/oreillymedia*

Watch us on YouTube: *http://www.youtube.com/oreillymedia*

Acknowledgments

It was a pleasure working on this project with a familiar group of friends and coworkers. Mike Loukides guided this project with his familiar gentle touch. Matt Darnell again did a tremendous job of copyediting the manuscript. The production team at O'Reilly was most accommodating when it came to my special requests regarding math typesetting and graphics.

Ben Peirce read the entire manuscript and provided valuable comments. I also acknowledge useful conversations with Austin King, Chris Nauroth, and Joe Adler.

I am especially indebted to Richard Kreckel, who carefully read several drafts of this manuscript and made many exceptionally valuable and insightful suggestions. I owe him big-time.

This book was written on Linux while running IceWM, tcsh, and XEmacs. The manuscript was prepared using LaTeX together with the AMS-LaTeX packages; the LaTeX manuscript was then transformed into the publisher's internal format for production. The graphs were drawn using gnuplot and pic; some calculations for the root locus diagrams in Chapter 24 were performed with Scilab. The simulations were implemented in Python.

PART I

Foundations

CHAPTER 1

Why Feedback? An Invitation

Workflow, order processing, ad delivery, supply chain management—enterprise systems are often built to maintain the *flow* of certain items through various processing steps. For instance, at a well-known online retailer, one of our systems was responsible for managing the flow of packages through the facilities. Our primary control mechanism was the number of pending orders we would release to the warehouses at any one time. Over time, these orders would turn into shipments and be ready to be loaded onto trucks. The big problem was to throttle the flow of pending orders just right so that the warehouses were never idle, but without overflowing them (quite literally) either.

Later I encountered exactly the same problem, but in an entirely different context, at a large publisher of Internet display ads. In this case, the flow consisted of ad impressions.[1] Again, the primary "knob" that we could adjust was the number of ads released to the web servers, but the constraint was a different one. Overflowing the servers was not a concern, but it was essential to achieve an even delivery of ads from various campaigns over the course of the month. Because the intensity of web traffic changes from hour to hour and from day to day, we were constantly struggling to accomplish this goal.

As these two examples demonstrate, maintaining an even flow of items or work units, while neither overwhelming nor starving downstream processing steps, is a common objective when building enterprise sys-

---

1. Every time an advertisement is shown on a website, this event is counted as an *impression*. The concept is important in the advertising industry, since advertisers often buy a certain number of such impressions.

tems. However, the changes and uncertainties that are present in all real-world processes frequently make it difficult, if not impossible, to achieve this goal. Conveyors run slower than expected and web traffic suddenly spikes, disrupting all carefully made plans. To succeed, we therefore require systems that can *detect* changes in the environment and *respond* to them.

In this book, we will study a particular strategy that has proven its effectiveness many times in all forms of engineering, but that has rarely been exploited in software development: feedback control. The essential ingredient is that we base the operations of our system specifically on the system's *output*, rather than on other, more general environmental factors. (For example, instead of monitoring the ups and downs of web traffic directly, we will base our delivery plan only on the actual rate at which ads are being served.) By taking the actual output into account (that's what "feedback" means), we establish a firm and reliable control over the system's behavior. At the same time, feedback introduces complexity and the risk of *instability*, which occurs when inappropriate control actions reinforce each other, and much of our attention will be devoted to techniques that prevent this problem. Once properly implemented, however, feedback control leads to systems that exhibit reliable behavior, even when subject to uncertainty and change.

A Hands-On Example

As we have seen, flow control is a common objective in enterprise systems. Unfortunately, things often seem rigged to make this objective difficult to attain. Here is a typical scenario (see Figure 1-1).

1. We are in charge of a system that releases items to a downstream processing step.
2. The downstream system maintains a buffer of items.
3. At each time step, the downstream system completes work on some number of items from its buffer. Completed items are removed from the buffer (and presumably kicked down to the next processing step).
4. We cannot put items directly into the downstream buffer. Instead, we can only release items into a "ready pool," from which they will eventually transfer into the downstream buffer.

5. Once we have placed items into the ready pool, we can no longer influence their fate: they will move into the downstream buffer owing to factors beyond our control.

6. The number of items that are completed by the downstream system (step 3) or that move from the ready pool to the downstream buffer (step 5) fluctuates randomly.

7. At each time step, we need to decide how many items to release into the ready pool in order to keep the downstream buffer filled without overflowing it. In fact, the owners of the downstream system would like us to keep the number of items in their buffer constant at all times.

Figure 1-1. Block diagram of a workflow system. Items are being released into the "ready pool," from which they are transferred to the downstream buffer.

It is somewhat natural at this point to say: this is unfair! We are supposed to control a quantity (the number of units in the downstream buffer) that we can't even manipulate directly. How are we supposed to do that—in particular, given that the downstream people can't even keep constant the number of items they complete at each time step? Unfortunately, life isn't always fair.

Hoping for the Best

What are we to do? One way of approaching this problem is to realize that, in the steady state, the number of units flowing *into* the buffer must equal the number of units flowing *out*. We can therefore measure the average number of units leaving the buffer at each time step and then make sure we release the same number of units into the ready pool. In the long run, things should just work out. Right?

Figure 1-2 (top) shows what happens when we do this. The number of units in the buffer (the queue length) fluctuates wildly—sometimes exceeding 100 units and other times dropping down to zero. If the space in the buffer is limited (which may well be the case if we are dealing with a physical processing plant), then we may frequently be overflowing the buffer. Even so, we cannot even always keep the downstream guys busy, since at times we can't prevent the buffer from running empty. But things may turn out even worse. Recall that we had to *measure* the rate at which the downstream system is completing orders. In Figure 1-3 (bottom) we see what happens to the buffer length if we underestimate the outflow rate by as little as 2 percent: We keep pushing more items downstream than can be processed, and it doesn't take long before the queue length "explodes." If you get paged every time this happens, finding a better solution becomes a priority.

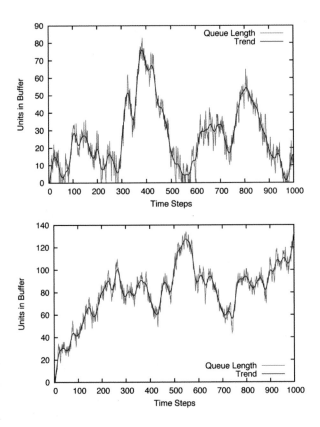

Figure 1-2. Number of units in the buffer as a function of time, where the consumption rate is equal to the inflow (top) or slightly smaller than the inflow (bottom).

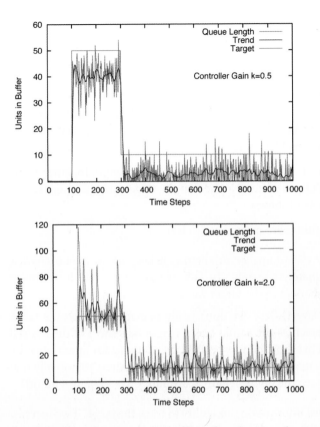

Figure 1-3. Number of units in the buffer, after the introduction of a controller that bases the inflow on the buffer's current fill level, for two different values of the controller gain k.

Establishing Control

Clearly, we need to come up with a better idea. The first step is to stop relying on the "average" outflow rate (which, by the way, may itself be changing as time goes on). Instead, we will monitor the *actual* length of the queue from moment to moment. In fact, we will ask the downstream team to give us a target: a specific queue length that they want us to maintain. We will then compare the actual queue length to the target. Only if the actual length is below the desired value will we release additional units into the ready pool. Moreover—and this is important—we will let the number of units released depend on the magnitude of the deviation: If the actual queue length is only slightly below the target, then we will release fewer units than if the queue length is way off the mark. Specifically, we will use the following formula to calculate the number of units to release into the ready pool:

released units = $k \cdot$ (target − actual)

where k is a numerical constant. How large should k be? Aye, there's the rub. We don't know yet. Why don't we take $k = 0.5$ for starters—that seems like a safe value.

Figure 1-3 (top) shows the actual queue length together with the desired target value. (Notice that the target value changes twice during the period shown.)

Two things are immediately clear:

- We are doing a *much* better job keeping the queue length approximately constant. In fact, we are even able to follow the two changes in the target value without too much trouble.

- Nevertheless, we don't really manage to match the target value exactly—the actual queue length falls short of the desired value. This shortfall is especially pronounced for later times, when the target itself is small: instead of the desired 10, we only manage to keep the queue length at around 3. That's quite a bit off!

Can we improve on our ability to track the target if we increase k (the "controller gain")? Figure 1-3 (bottom) shows what happens if we set $k = 2.0$. Now the actual queue length *does* match the target in the long run, but the queue length is fluctuating a great deal. In particular, when the system is first switched on we overshoot by more than a factor of 2. That's probably not acceptable.

In short, we are clearly on a good path. After all, the behavior shown in Figure 1-3 is incomparably better than what we started with (Figure 1-2). Yet we are probably still not ready for prime time.

Adding It Up

If we reflect on the way our control strategy works, we can see where it falls short: we based the corrective action (that is, the number of units to be released into the ready pool) on the magnitude of the deviation from the target value. The problem with this procedure is that, if the deviation is small, then the corrective action is also small. For instance, if we set $k = 0.5$, choose 50 as a target, and the current length of the queue is 40, the corrective action is $0.5 \cdot (50 - 40) = 5$. This happens to be approximately the number of units that are removed from the buffer by the downstream process, so we are never able to bring the queue length up to the desired value. We can overcome this problem by increasing the controller gain, but with the result that then we will occasionally overshoot by an unacceptable amount.

The remedy is to magnify the effect of persistent small deviations by adding them up! After a few time steps, their influence will have grown sufficiently to make itself felt. However, if the deviations are reliably zero, then adding them up does not make a difference. Hence we modify our control strategy as follows. We still calculate the tracking error as (target – actual) at each time step, but we also keep a running sum of *all* tracking errors up to this point. We then calculate the number of units to be released as the combination of the two contributions:

released units $= k_p \cdot$ error $+ k_i \cdot$ cumulative error

Now we have two numerical factors to worry about: one (k_p) for the term that is proportional to the error, and one (k_i) for the term that is proportional to the cumulative sum (or: the "integral") of the error. After some trial and error, we can obtain a result that's quite adequate (see Figure 1-4).

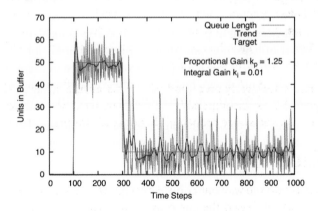

Figure 1-4. The number of units in the buffer when using a controller that contains both a proportional and an integral term.

Summary

The method we have utilized in this chapter is called *feedback*. The goal of using feedback control is to make a system's output track a reference signal as closely as possible. This is achieved by continuously comparing the output signal to the reference and applying a corrective action to reduce the tracking error. Moreover, the magnitude of the corrective action depends on the magnitude of the tracking error.

Feedback is a fairly robust control strategy—using feedback, one can successfully track a reference signal even in the face of uncertainty. The reason for the uncertainty does not matter: it may be due to random effects ("noise," as in our queueing example); it may be due to our lack of knowledge about the inner workings of a complicated, "black-box" system that we need to control. In contrast, *feedforward control*—whereby we attempt to work out all decisions ahead of time—requires precise knowledge of all applicable laws yet still remains vulnerable to the detrimental effects of inaccuracy and uncertainty.

The behavior of systems involving feedback loops can be complicated and hard to predict intuitively. In particular, we run the danger of introducing *instability*: a corrective action leads to overshoot, which in turn is corrected by an even greater overshoot in the opposite direction, in a rapidly escalating pattern. Even if such catastrophic blow-ups can be avoided, it is often the case that feedback-controlled systems exhibit undesirable oscillatory behavior (*control oscillations*).

The desire to avoid instability in feedback loops has led to the development of a deep and impressive theory describing such systems but also to a set of heuristics and simple "rules of thumb" for practical applications. Part II of this book provides an overview of some of these heuristic tuning and design methods, and Part IV offers a brief introduction to the central theoretical methods and results.

Let's close this first look with some general observations about feedback control and the types of situations where it is most applicable.

- Feedback control applies an automatic correction to deviations from a reference signal. This allows for tighter control but is accompanied by a tendency toward oscillatory or even unstable behavior.

- Feedback is about tracking a given reference signal. Without a reference signal, there can be no feedback control.

- It follows that feedback is about *control*, not about *optimization*. A task such as "make the flow through the system as large as possible" can not be solved by feedback alone—instead, such a task requires an optimization strategy. However, feedback may prove extremely useful if not necessary when you are seeking to *implement* or execute such a strategy.

- Frequent, small changes are better suited to stabilizing a system than rare, large changes. If we are unable to observe the system constantly and apply corrective actions frequently, then feedback won't work.

- The choice of reference signal will often be determined by an existing optimization strategy. In contrast, what is considered to be a system's "input" and "output" is arbitrary and depends on the objectives and any existing constraints. In our example, we tried to maintain a constant queue length in the buffer. Alternatively, we could have tried, for example, to maintain a certain throughput through the entire system. Identifying the most suitable input/output variables to accomplish the desired task most easily can be a real challenge, especially when applying feedback control to a "nonstandard" situation.

Code to Play With

The graphs in Figure 1-2 through Figure 1-4 were produced using a simulation of the buffer system described. The simulation is simple, so it is easy to add your own extensions and variations. Play around with this system a little: make some changes and see how it affects the outcomes. It is *extremely* helpful to develop some experience with (and intuition for) the effects that feedback control can have!

We begin with the class for the entire buffer system, including what we have called the "ready pool."

```python
class Buffer:
    def __init__( self, max_wip, max_flow ):
        self.queued = 0
        self.wip = 0        # work-in-progress ("ready pool")

        self.max_wip = max_wip
        self.max_flow = max_flow # avg outflow is max_flow/2

    def work( self, u ):
        # Add to ready pool
```

```
u = max( 0, int(round(u)) )
u = min( u, self.max_wip )
self.wip += u

# Transfer from ready pool to queue
r = int( round( random.uniform( 0, self.wip ) ) )
self.wip -= r
self.queued += r

# Release from queue to downstream process
r = int( round( random.uniform( 0, self.max_flow ) ) )
r = min( r, self.queued )
self.queued -= r

return self.queued
```

The Buffer maintains two pieces of state: the number of items currently in the buffer, and the number of items in the ready pool. At each time step, the work() function is called, taking the number of units to be added to the ready pool as its argument. Some constraints and business rules are applied; for example, the number of units must be a positive integer, and the number of items added each time is limited—presumably by some physical constraint of the real production line.

Then, a random fraction of the ready pool is promoted to the actual buffer. (This could be modeled differently—for instance by taking into account the amount of time each unit has spent in the ready pool already.) Finally, a random number of units is "completed" at each time step and leaves the buffer, with the average number of units completed at each time step being a constant.

Compared with the downstream system, the Controller is much simpler.

```
class Controller:
    def __init__( self, kp, ki ):
        self.kp, self.ki = kp, ki
        self.i = 0        # Cumulative error ("integral")

    def work( self, e ):
        self.i += e

        return self.kp*e + self.ki*self.i
```

The Controller is configured with two numerical parameters for the controller gain. At each time step, the controller is passed the tracking error as argument. It keeps the cumulative sum of all errors and produces a corrective action based on the current and the accumulated error.

Finally, we need two driver functions that utilize those classes: one for the open-loop mode of Figure 1-2 and one for the closed-loop operation of Figure 1-3 and Figure 1-4.

```
def open_loop( p, tm=5000 ):
    def target( t ):
        return 5.0  # 5.1

    for t in range( tm ):
        u = target(t)
        y = p.work( u )

        print t, u, 0, u, y

def closed_loop( c, p, tm=5000 ):
    def setpoint( t ):
        if t < 100: return 0
        if t < 300: return 50
        return 10

    y = 0
    for t in range( tm ):
        r = setpoint(t)
        e = r - y
        u = c.work(e)
        y = p.work(u)

        print t, r, e, u, y
```

The functions take the system or process p (in our case, an instance of Buffer) and an instance of the Controller c (for closed-loop operations) as well as a maximum number of simulation steps. Both functions define a nested function to provide the momentary target value and then simply run the simulation, printing progress to standard output. (The results can be graphed any way you like. One option is to use gnuplot—see Appendix B for a brief tutorial.)

With these class and function definitions in place, a simulation run is easy to undertake:

```
import random

class Queue:
    ...

class Controller:
    ...
```

```python
def open_loop():
    ...

def closed_loop():
    ...


c = Controller( 1.25, 0.01 )
p = Queue()

open_loop( p, 1000 )

# or: closed_loop( c, p, 1000 )
```

CHAPTER 2

Feedback Systems

The method we employed in the previous chapter was based on the *feedback principle*. Its basic idea can be simply stated as follows.

Feedback Principle

Continuously compare the actual output to its desired reference value; then apply a change to the system inputs that counteracts any deviation of the actual output from the reference.

In other words, if the output is too high, then apply a correction to the input that will lead to a reduction in output; if the output is below the reference, then apply a correction to the input that raises the value of the output.

The essential idea utilized by the feedback concept is to "loop the system output back" and use it for the calculation of the input. This leads to the generic feedback or *closed-loop* architecture (see Figure 2-1). This should be compared to the feedforward or *open-loop* architecture (Figure 2-2), which does not take the system output into account.

Basing the calculation of the next input value on the previous output implies that feedback is an *iterative* scheme. Each control action is intended only to take the system *closer* to the desired value, which it does by making a step in the right direction. No special effort is made to eliminate the difference between reference and output entirely; instead, we rely on the *repetition* of steps that merely reduce the error.

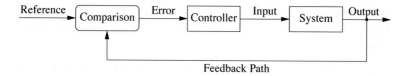

Figure 2-1. *The structure of a feedback loop: the system's output is routed back and compared to the reference value in order to calculate a new input to the system.*

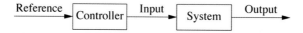

Figure 2-2. *The structure of an open-loop (or feedforward) arrangement: the system input is calculated directly from the reference value without taking the system output into account.*

As with any iterative scheme, three questions present themselves:

- Does the iteration *converge*? (Or does it diverge?)
- How *quickly* does it converge? (If it converges at all.)
- What value does it converge *to*? (Does it converge to the desired solution or to a different one?)

Systems and Signals

We will consider different systems in this book that serve a variety of purposes. What these systems have in common is that all of them depend on *configuration* or *tuning parameters* that affect the system's behavior. To obtain knowledge about that behavior, we track or observe various *monitored metrics*. In most cases, the system is expected to meet or exceed some predefined *quality-of-service* measure. The *control problem* therefore consists of adjusting the configuration parameters in such a way that the monitored metrics fall within the range prescribed by the quality-of-service requirements.

As far as the control problem is concerned, the configuration parameters are the variables that we can influence or manipulate directly. They are sometimes called the *manipulated variables* or simply the (control) *inputs*. The monitored metrics are the variables that we want to influence, and they are occasionally known as the *process variables* or the (control) *outputs*. The inputs and outputs taken together constitute the *control signals*.

The terms "input" and "output" for (respectively) the manipulated and the tracked quantities, are very handy, and we will use them often. However, it is important to keep in mind that this terminology refers only to the purpose of those quantities *with respect to the control problem* and so has nothing to do with functional "inputs" or "outputs" of the system. Whenever there is any risk of confusion, use the terms "configurable parameter" and "tracked metric" in place of "input" and "output."

For the most part we will consider only those systems that have *exactly one* control input and control output, so that there is only a single configurable parameter that can be adjusted in order to influence the a single tracked metric. Although this may seem like an extreme limitation, it does cover a wide variety of systems. (Treating systems that have multiple inputs or outputs is possible in principle, but it poses serious practical problems.)

Here is a list of some systems and their input and output signals from enterprise programming and software engineering.

A cache:
> The tracked metric is the hit rate, and the configurable variable is the cache size (the maximum number of items that the cache can hold).[1]

A server farm:
> The tracked metric is the response latency, and the adjustable parameter is the number of servers online.

A queueing system:
> The tracked metric is the waiting time, and the adjustable parameter is the number of workers serving the queue.

A graphics library:
> The tracked quantity is the total amount of memory consumed, and the configurable quantity is the resolution.

---

1. The manipulated parameter (and hence the input from a controls perspective) is the cache size. This must not be confused with the functional "input" of cached items to the cache.

Other examples:

A heated room or vessel:
> The tracked metric is the temperature in the room or vessel, and the adjustable quantity is the amount of heat supplied. (For a pot on the stove, the adjustable quantity is the setting on the dial.)

A CPU cooler:
> The tracked metric is the CPU temperature, and the adjustable quantity is the voltage applied to the fan.

Cruise control in a car:
> The tracked metric is the car's speed, and the adjustable quantity is the accelerator setting.

A sales situation:
> The tracked metric is the number of units sold, and the adjustable quantity is the price per item.

Tracking Error and Corrective Action

The feedback principle demands that the process output be constantly compared to the reference value (usually known as the *setpoint*). The deviation of the actual process output from the setpoint is the *tracking error*:

tracking error = setpoint − output

The job of the controller in Figure 2-1 is to calculate a corrective action based on the value of the tracking error. If the tracking error is positive (meaning that the process output is too low) then the controller must produce a new control input that will *raise* the output of the process, and vice versa.

Observe that the controller can do this without detailed knowledge about the system and its behavior. The controller mainly needs knowledge about the *directionality* of the process: does the input need to be increased or decreased in order to raise the output value? Both situations do occur: increasing the power supplied to a heating element will lead to an increase in temperature, but increasing the power supplied to a cooler will lead to a decrease!

Once the direction for the control action has been determined, the controller must also choose the *magnitude* of the correction. We will have more to say on this in the next section.

Stability, Performance, Accuracy

The introduction of a feedback loop can make an originally stable system unstable. The problem is usually due to persistent *overcompensation*, which results from corrective actions that are too large. Consider the cache (from the examples listed previously) and assume that the hit rate is initially *below* the desired value. To increase the hit rate, we need to make the cache larger. But how much larger? If we make the cache too large, then the hit rate will end up being *above* the desired value so that the cache size ends up being reduced in the next step; and so on. The system undergoes *control oscillations*, switching rapidly and violently between different configurations; see Figure 2-3.

Control oscillations are rarely desirable—just imagine the cruise control in your car behaving this way! But things can get worse: if each over- or undershoot leads to an even *larger* compensating action, then the amplitude of the oscillations grows with time. The system has thus become *unstable* and will break up (or blow up) before long. Such unstable behavior must be avoided in control loops at all cost.

The opposite problem is slow or sluggish behavior. If we are too timid and apply control actions that are too small, then the system will be slow to respond to disturbances and tracking errors will persist for a long time (Figure 2-3). Although less dangerous than instability, such sluggish behavior is also unsatisfactory. To achieve the quickest response, we will therefore want to apply the *largest control action that does not make the system unstable.*

A well-designed control system should show good *performance*, which means that it responds to changes quickly so that deviations between the tracked metric and the reference value do not persist. The typical response time of a control system describes how quickly it can react to changes and therefore establishes a limit on the fastest possible disturbances it will be able to handle.

In the steady state, the quality of a control system is measured by the *accuracy* with which it is able to follow a given reference value. The behavior of feedback control systems is usually evaluated in terms of stability, performance, and accuracy.

We can now recast our three questions about the convergence of an iterative system in these control-theoretic terms as follows.

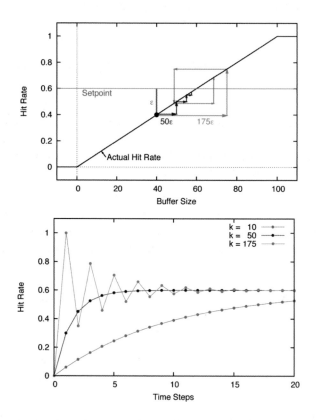

Figure 2-3. Finding the optimal size of a cache to achieve a desired hit rate. The top panel shows how an initial error leads to different iterations depending on the gain factor; the bottom panel shows the time evolution of the hit rate for different controller gains. Corrections that are too large lead to oscillatory behavior; corrections that are too small result in sluggish performance.

Stability:

Is the system stable? Does it respond to changes without undue oscillations? Is it guaranteed that the amplitude of oscillations will never build up over time instead will decay rapidly?

Performance:
> How quickly does the system respond to changes? Is it able to respond quickly enough for the given application? (The autopilot for a plane needs to respond faster than one for a ship.)

Accuracy:
> Does the system track the specified reference value with sufficient accuracy?

It turns out that not all of these goals can be achieved simultaneously. In particular, the design of a feedback system always involves a trade-off between stability and performance, because a system that responds quickly will tend to oscillate. Depending on the situation, one must choose which aspect to emphasize.

In general, it is better to make many small adjustments quickly than to make few large adjustments occasionally. With many small steps, a correcting action will be taken quickly—before the system has had much opportunity to build up a significant deviation from the desired value. If corrective actions are taken only rarely, then the magnitude of that deviation will be larger, which means there is a greater chance of overcompensation and the associated risks for instability.

The Setpoint

The purpose of feedback systems is to track a reference value or *setpoint*. The existence of such a reference value is mandatory; you can't have feedback control without a setpoint.

On the one hand, this is a triviality: there is obviously a desirable value for the tracked metric, for why would we track it otherwise? But one needs to understand the specific restrictive nature of the setpoint in feedback control.

By construction, a feedback loop will attempt to replicate the given reference value *exactly*. This rules out two other possible goals for a control system. A standard feedback loop is not suitable for maintaining a metric within a *range* of values; instead, it will try to drive the output metric to the precise value defined by the setpoint. For many applications, this is too rigid. Feedback systems require additional provisions to allow for floating range control (for instance, see Chapter 18).

Moreover, one must take care not to confuse feedback control with any form of *optimization*. Feedback control tries to replicate a setpoint, but it involves no notion of achieving the "best" or "optimal" output under a given set of conditions. That being said, a feedback system may well be an important part of an overall optimization strategy: if there is an overall optimization plan that prescribes what output value the system should maintain, then feedback control is the appropriate tool to deliver or execute this plan. But feedback control itself is not capable of identifying the optimal plan or setting.

One occasionally encounters additional challenges. For instance, the self-correcting nature of the feedback principle requires that the actual system output must be able to straddle the setpoint. Only if the output can fall on either side of the setpoint is the feedback system capable of applying a restoring action in either direction. For setpoint values at the end of the achievable range, this is not possible. Consider, for example, a cache. If we want the hit rate (as the tracked metric) to equal 100 percent, then the actual hit rate can never be greater than the setpoint; this renders impossible a corrective action that would diminish the size of the cache. (In Chapter 15 we will see some ad hoc measures that can be brought to bear in a comparable situation.)

Uncertainty and Change

Feedback systems are clearly more complicated than straightforward *feedforward* systems that do not involve feedback. The design of feedback loops requires balancing a variety of different properties and sometimes difficult trade-offs. Moreover, feedback systems introduce the risk of instability into otherwise stable systems and therefore necessitate extra measures to prevent "blow-ups." Given all these challenges, when and why are feedback systems worth the extra complexity? The answer is that feedback systems offer a way to achieve reliable behavior even *in the presence of uncertainty and change.*

The way configuration parameters (control inputs) affect the behavior of tracked metrics (control outputs) is not always well known. Consider the cache, again: making the cache larger will certainly increase the hit rate—but by how much? Just how large does the cache have to be in order to attain a specific hit rate? These are difficult questions whose answers depend strongly on the nature of the access patterns for cache items. (How many distinct items are being requested over some time period, and so on.) This ignorance regarding the relationship between inputs and outputs leads to *uncertainty.* But even if we were able to work out the input/output relation precisely at some particular time, the system would still be subject to *change*: the access patterns may (and will!) change over time. Different items are being requested. The distribution of item requests is different in the morning than in the afternoon. And so on.

Feedback is an appropriate mechanism to deal with these forms of uncertainty and change. In the absence of either factor, feedback would be unnecessary: if we know exactly how the cache size will affect the hit rate *and* if we know that access patterns are not subject to change, then there is no need for a feedback loop. Instead, we could simply choose the appropriate cache size and be done with it. But how often are we in that position?

To be fair, such situations do exist, mostly in isolated environments (and are thus not subject to change) with clearly defined, well-known rules (thereby avoiding uncertainty). Computer programs, for instance, are about as deterministic as it gets! Not much need for feedback control.

But the same cannot be said about computer *systems*. As soon as several components interact with one another, there is the possibility of randomness, uncertainty, and change. And once we throw human users into the mix, things can get pretty crazy. All of a sudden, uncertainty is guaranteed and change is constant. Hence the need for feedback control.

Feedback and Feedforward

One can certainly build feedforward systems intended to deal with complex and changing situations. Such systems will be relatively complicated, possibly requiring deep and complicated analysis of the laws governing the controlled system (such as understanding the nature of cache request traffic). Nevertheless, they may still prove unreliable—especially if something unexpected happens.

Feedback systems take a very different approach. They are intentionally simple and require only minimal knowledge about the controlled system. Only two pieces of information are really needed: the *direction* of the relationship between input and output (does increasing the input drive the output up or down?) and the *approximate* magnitude of the quantities involved. Instead of relying on detailed understanding of the controlled system, feedback systems rely on the ability to apply corrective actions repeatedly and quickly.

In contrast to a typical feedforward system, the iterative nature of feedback systems makes them, in some sense, nondeterministic. Instead of mapping out a global plan, they only calculate a local change and rely on repetition to drive the system to the desired behavior. At the same time, it is precisely the *absence* of a global plan that allows these systems to perform well in situations characterized by uncertainty and change.

Feedback and Enterprise Systems

Enterprise systems (order or workflow processing systems, request handlers, messaging infrastructure, and so on) are complicated, with many interconnected but independently operating parts. They are connected to the outside world and hence are subject to change, which may be periodic (hour of the day, day of the week) or truly random.

In my experience, it is customary to steer such enterprise systems using feedforward ideas. To cope with the inevitable complexities, programmers and administrators resort to an ever-growing multitude of "configurational parameters" that control flow rates, active server instances, bucket sizes, or what have you. Purely numerical weighting factors or multipliers are common. These need to be adjusted manually in order to accommodate changing conditions or to optimize the systems in some way, and the effect of these adjustments is often difficult to predict.[2] There may even be complete subsystems that change the values of these parameters according to the time of day or some other schedule—but still in a strictly feedforward manner.

I believe that feedback control is an attractive alternative to all that, and one that has yet to be explored. Enjoy!

Code to Play With

The following brief program demonstrates the effect that the magnitude of the corrective action has on the speed and nature of the iteration. Assume that our intent is to control the size of a cache so that

---

2. Shortly before I began preparing this book, I overheard a coworker explain to a colleague: "The database field 'priority' here is used by the delivery system to sort different contracts. It typically is set by the account manager to some value between 0.1 and 10 billion." There *has* to be a better way!

the success or hit rate for cache requests is 60 percent. Clearly, making the cache larger (smaller) will increase (decrease) the hit rate.

In the code that follows we do not actually implement a cache (we will do so in Chapter 13), we just mock one up. The function cache() takes the size of the cache and returns the resulting hit rate, which is implemented as size/100. If the size falls below 0 or grows over 100, then the reported hit rate is (respectively) 0 or 1. (Of course, other relations between size and hit rate are possible—feel free to experiment.)

The program reads a gain factor k from the command line. This factor controls the size of the corrective action that is being applied during the iteration. The iteration itself first calculates the tracking error e as the difference between setpoint r and actual hit rate y; it then calculates the cumulative tracking error c. The new cache size is computed as the product of the cumulative error and the gain factor: k*c.[3]

Depending on the value provided for the gain factor k, the iteration will converge to the steady-state value more or less quickly and will either oscillate or not. It will never *diverge* completely because the system output is constrained to lie between 0 and 1; hence the output cannot grow beyond all bounds. Examples of the observable behavior are shown in the top panel of Figure 2-3.

```
import sys
import math

r = 0.6                    # reference value or "setpoint"
k = float( sys.argv[1] ) # gain factor: 50..175

print "r=%f\tk=%f\n" % ( r, k ); t=0
print r, 0, 0, 0, 0

def cache( size ):
    if size < 0:
        hitrate = 0
    elif size > 100:
        hitrate = 1
    else:
        hitrate = size/100.0
    return hitrate
```

---

3. In case you are wondering why the updating is based on the cumulative error c instead of the tracking error e, feel free to experiment. Use u = k*e instead and observe the results closely. Can you see what is going on and why? The solution will be given (and discussed in detail) in Chapter 4.

```
y, c = 0, 0
for _ in range( 200 ):
    e = r - y        # tracking error
    c += e           # cumulative error
    u = k*c          # control action: cache size
    y = cache(u)     # process output: hitrate

    print r, e, c, u, y
```

CHAPTER 3

System Dynamics

In Chapter 2 we discussed how a feedback loop can drive the output of a system to a desired value. We also described the typical challenges associated with this scheme: making sure that the overall system is stable (meaning that it does, in fact, converge to the preset value) and performs well (so that it converges quickly). However, this is not the whole story.

Lags and Delays

The aspect that we have neglected so far is that many systems do not respond *immediately* to a control input; instead, they respond with some form of lag or delay. In addition, many systems exhibit even more complicated behavior when stimulated. These factors need to be taken into account when designing a control loop.

Let's consider a few examples (see Figure 3-1 and Figure 3-2).

A heated vessel:
> (Basically, this describes a pot on the stove.) As the heat is turned on, the temperature in the vessel does not immediately jump to its final value; rather, it shows a gradual response. Likewise, when the external heat supply is later shut off, the temperature in the vessel does not drop immediately but instead slowly decays to the ambient temperature. The temperature in the vessel (which is the tracked quantity or "output") follows the temperature (which is the configurable quantity or "input"), but it lags behind and shows a more "rounded" behavior. This form of rounded, partial response is called a *lag*.

A tank fed by a pipe:

Imagine a long hose or pipe that feeds into a storage tank. When the valve at the beginning of the pipe is opened, the level in the tank (which is the tracked metric in this case) does not change. Only after the liquid has traveled the entire distance of the pipeline and begins to flow into the tank does the tank's fill level begin to rise. This type of behavior is called a *delay*. In contrast to a lag, which consists of an immediate but partial response, a delay is characterized by an initial time interval during which there is no response at all.

A mass on a spring:

Consider a mass supported by a spring. If we give the other end of the spring a sudden jerk, the mass on the spring will begin to oscillate. This system exhibits a lag, because there is an immediate but gradual response. In addition, however, this system also displays complicated behavior on its own; it has nontrivial *internal dynamics*.

A fishing rod:

When a flexible fishing rod is yanked *back*, its tip initially moves *forward*. In other words, for this system the first response to an external stimulus is in the direction *opposite* to the stimulus. Such systems occasionally do occur in practice and obviously pose particular challenges for any controller. They are known as *non-minimum phase* systems (for reasons involving the theoretical description of such behavior), but a more descriptive name would be "inverse response" systems.

Of course, all of these behaviors can also occur in combination.

Forced Response and Free Response

The examples discussed so far also demonstrate the difference between *forced* and *free* (or internal) response. In the case of the heated pot, the steady increase in temperature while the stove is turned on is the response to the external disturbance (namely, the applied heat). In contrast, the exponential cooling off is subject only to the system's internal structure (because there is no input being applied from the outside). The initial displacement of the mass on the spring is due to the externally applied force on the free end of the spring; the subsequent oscillation is the free response of the spring–mass system itself. The storage tank does not have any free response: its fill level does not change unless liquid is flowing in. The dynamic response of the fishing rod is extremely complicated and includes both forced and free components.

Transient Response and Steady-State Response

The response of a system to an external disturbance often consists of both a *transient* component, which disappears over time, and a *steady-state* component. These components are well illustrated by the spring–mass system: the initial oscillations die away in time, so they are transient. But the overall displacement of the mass, which is the result of the change in position of the spring's free end, persists in the steady state.

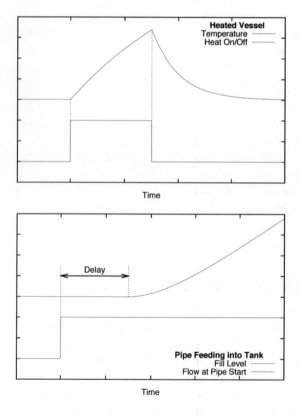

Figure 3-1. Examples of dynamical systems and their response to an external stimulus. These types of behavior are common among industrial processes and computer systems.

Control actions are often applied to bring about a change in the steady-state output. (If you turn up the heat, you want the room to be warmer and to stay that way.) From this perspective, transient responses are viewed mainly as a nuisance: unwanted accompaniments of an applied change. Hence an important measure for the performance of a control system is the time it takes for all transient components of the response to have disappeared. Usually, handling the transient response in a control system involves an engineering trade-off: systems in which transients are strongly damped (so that they disappear quickly) will respond more slowly to control inputs than do systems in which transient behavior is suppressed less strongly.

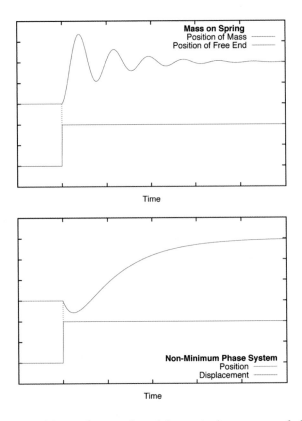

Figure 3-2. Additional examples of dynamical systems and their response to an external stimulus. The oscillatory response is especially important among mechanical and electrical systems. The initial response in the direction opposite to the external influence makes non-minimum phase systems difficult to control.

Dynamics in the Physical World and in the Virtual World

All objects in the physical world exhibit some form of lag or delay, and most mechanical or electrical systems will also have a tendency to oscillate.

The reason for the lags or delays is that the world is (to use the proper mathematical term) *continuous*: An object that is in some place now cannot be at some totally different place a moment later. The object has to move, continuously, from its initial to its final position. It can't move infinitely quickly, either. Moreover, to move a physical object very quickly requires large amounts of force, energy, and power (in the physical sense), which may not be available or may even be impossible to supply. (Accelerating a car from 0 to 100 km/h in 5 seconds takes a large engine, to do so in 0.5 seconds would take a significantly larger engine, and a very different mode of propulsion, too.) The amount by which objects in the physical world can move (or change their state in any other way) is limited by the laws of nature. These laws rule supreme: No amount of technical trickery can circumvent them.

For computer systems, however, these limitations do not necessarily apply in quite this way! A computer *program* can arbitrarily change its internal state. If I want to adjust the cache size from 10 items to 10 billion items, there is nothing to stop me—the next time through the loop, the cache will have the new size. (The cache size may be limited by the amount of memory available, but that is not a fundamental limitation—you can always buy some more.) In particular, computer programs do not exhibit the partial response that is typical of continuous systems. You will typically get the entire response at once. If you asked for a 100-item cache, you will get the entire lot the next time through the loop—not one item now, five more the next time around, another 20 coming later, and so on. In other words, computer programs typically do not exhibit lags. They do, however, exhibit delays; these are referred to as "latency" in computer terminology. If you are asking for 20 additional server instances in your cloud data center, it will take a few minutes to spin them up. During that time, they are not available to handle requests—not even partially. But once online, they are immediately fully operational. Figure 3-3 shows schematically the typical response characteristic of a computer program. Compared to the "smooth" behavior of physical objects displayed in Figure 3-1 and

Figure 3-2, the dynamics of computer applications tend to be discontinuous and "hard."

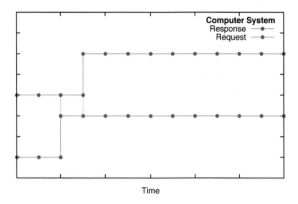

Figure 3-3. Typical dynamical response of a computer system: The response is discontinuous and occurs after a measurable delay.

The considerations of the preceding paragraph apply to standalone computer applications. For computer *systems*, which consist of multiple computers (or computer programs) communicating with each other and perhaps with the rest of the world (including human users), things are not quite as clear-cut. For instance, recall our example of the item cache. We can adjust the *size* of the cache from one iteration to the next, and by an arbitrary amount. But it does not follow that the *hit rate* (the tracked quantity or "output" for this system) will respond just as quickly. To the contrary, the hit rate will show a relatively smooth, lag-type response. The reason is that it takes time for the cache to load, which slows the response down. Moreover, the hit rate itself is necessarily calculated as some form of trailing average over recent requests, so some time must elapse before the new cache size makes itself felt. In the case of the additional server instances being brought up in the data center, we may find that not all the requested instances come online at precisely the same moment but instead one after another. This also will tend to "round out" the observed response.

Another case of computer systems exhibiting nontrivial dynamics consists of systems that already include a "controller" (or control algorithm) of some form. For example, it is common practice to *double* the size of a buffer whenever one runs out of space. Similar strategies can be found, for instance, in network protocols, for the purpose of maximizing throughput without causing congestion. In these and similar situations, the behavior of the computer system is constrained by its own internal control algorithms and will not change in arbitrary ways.

Dynamics and Memory

We saw that objects in the physical world cannot suddenly change their state (their position, temperature, whatever): Changes must occur continuously. This is another way of saying that the state of such an object is not independent of its past. These objects have a "memory" of their past, and it is this memory that leads to nontrivial dynamics. To make this point more concrete: The pot on the stove does not just "forget" its current temperature when the stove is turned on; the pot "remembers" its original temperature and therefore takes time to adjust.

For objects in the physical world, different modes of energy storage form the mechanism for the kind of memory that leads to "lags"; in contrast "delays" are generally due to transport phenomena.

For computer systems, we need to evaluate whether or not they do possess a "memory" (in the sense discussed here). The cache *size* does not have a memory: It can change immediately and by an arbitrary amount. The cache's *hit rate*, however, does retain knowledge of its past—not only through the cache loading but also through the averaging process implicit in the calculation of the hit rate.

All mechanisms that explicitly retain knowledge of their past are likely to give rise to lags. With computer systems, words such as "buffer," "cache," and "queue" serve as indicators for nonimmediate responses. Another source of "memory" is any form of "time averaging," "filtering," or "smoothing." All of these operations involve the current value of some quantity as well as past values, thus leading to nontrivial dynamical behavior. Lags tend to be collective phenomena.

Delays (or "latency") tend to result from transport issues (as in the physical world—think of network traffic) or to internal processes of the system that are not observable in the monitored metrics. Examples include the boot-up process of newly commissioned server instances, the processing time of database requests, and also the delays that individual events might experience while stuck in some form of opaque queue. Delays can occur both for individual systems and for events.

The Importance of Lags and Delays for Feedback Loops

The reason we spend so much time discussing these topics is that lags and delays make it much more difficult to design a control system that is both stable and performant. Take the heated vessel: Initially, we want to raise its temperature and so we apply some heat. When we check the output a short while later, we find that the temperature has barely moved (because it's lagging behind). If we now increase the heat input, then we will end up overheating the vessel. To avoid this outcome, we must take the presence (and length) of the lag into account.

Moreover, unless we are careful we might find ourselves applying corrective action at precisely the wrong moment: Applying a control action intended to reduce the output precisely when the output has already begun to diminish (but before this is apparent in the actual system output). If we reach this scenario, then the closed-loop system undergoes sustained oscillations. (Toward the end of this chapter, you will find a brief computer exercise that demonstrates how the delay of even a single step can lead to precisely this situation.)

Avoiding Delays

The whole feedback principle is based on the idea of applying corrective actions in response to deviations of the output from the reference value. This scheme works better the more quickly any deviation is detected: If we detect a deviation early, before it has had a chance to become large, then the corrective action can be small. This is good not only because it is obviously desirable to keep the tracking error small (which is, after all, the whole point of the exercise!) but also because small control actions are more likely to lead to stable behavior. Furthermore, there is often a "cost" associated with control actions, making large movements more expensive—in terms of wear and tear, for instance. (This is not always true, however: sometimes there is a fixed cost associated with making a control action, independent of its size. When adding server instances, for example, there may be a fixed commissioning fee in addition to the cost for CPU time used. In such cases, we naturally want to limit the number of control actions. Nevertheless, early detection of deviations is still relevant. It is up to the controller to decide how to react.)

For this reason, we should make an effort to ensure timely observation of all relevant quantities. For physical systems, this means using quick-acting sensors and placing them close to the action. Old-style thermometers, for instance, have their own lags (they are nothing but "heated vessels" as discussed earlier) and transport delays (if they are placed far away from the heat source).

Because computer systems manage "logical" signals, there is often greater freedom in the choice of quantities that we use as monitoring or "output" signals than for systems in the physical world. To quantify the performance of a server farm, we can use the number of requests pending, the average (or maximum) age of all requests, the number (or fraction) of dropped requests, the arrival rate of incoming request, the response time, and several other metrics. We should make the best use of this freedom and make an effort to identify and use those output variables that respond the most quickly to changes in process input. In general, metrics that are calculated as averages or other summaries will tend to respond more slowly than quantities based on individual observations; the same is true for quantities that have been "smoothed" to avoid noise. In fact, it is often better to use a noisy signal directly than to run it through a smoothing filter: The slowdown incurred through the filtering outweighs the benefits of having a smoother signal. We will discuss some relevant choices in Chapter 5.

There is usually little that can be done about lags and delays inherent in the dynamics of the controlled system—simply because the system is not open to modification. (But don't rule this possibility out, especially if the "system" is merely a computer program rather than, say, a chemical plant.) We should, however, make every effort to avoid delays in the architecture of the control loop. Chapter 15 and Chapter 16 will provide some examples.

Theory and Practice

There exists a very well-developed, beautiful, and rather deep theory to describe the dynamic behavior of feedback loops, which we will sketch in Part IV. An essential ingredient of this theory is knowledge of the dynamic behavior of the controlled system. If we can describe how the controlled system behaves *by itself*, then the theory helps us understand how it will behave as part of a feedback loop. In particular, the theory is useful for understanding and calculating the three essential properties of a feedback system (stability, performance, and accuracy) even in the presence of lags and delays.

The theory does require a reasonably accurate description of the dynamics of the system that we wish to control. For many systems in the physical world, such descriptions are available in the form of differential equations. In particular, simple equations describing mechanical, electrical, and thermal systems are well known (the so-called laws of nature), and the entire theory of feedback systems was really conceived with them in mind.

For computer systems, this is not the case. There are no laws, outside the program itself, that govern the behavior of a computer program. For entire computer systems there may be applicable laws, but they are neither simple nor universal; furthermore, these laws are not known with anything like the degree of certainty that applies to, say, a mechanical assembly. For instance, one can (at least in principle) use methods from the theory of stochastic processes to work out how long it will take for a cache to repopulate after it has been resized. But such results are difficult to obtain, are likely to be only approximate, and in any case depend critically on the nature of the traffic—which itself is probably not known precisely.

This does not mean that feedback methods are not applicable to computer systems—they are! But it does mean that the existing theory is less easily applied and provides less help and insight than one might wish. Developing an equivalent body of theoretical understanding for computer systems and their dynamics is a research job for the future.

Code to Play With

To understand how even a simple delay can give rise to nontrivial behavior when encountered in a feedback architecture, let's consider a system that simply replicates the input from the *previous* time step to its output. We close the feedback loop as in Figure 2-1 and use a controller, which merely multiplies its input by some constant k.

The brief listing that follows shows a program that can be used to experiment with this closed-loop system. The program reads both the setpoint r and the controller gain k from the command line. The iteration itself is simple: The tracking error and the controller output are calculated as in Chapter 2, but the current output y is set to the value of the controller output from the *previous* time step. Figure 3-4 shows

the time evolution of the output y for two different values of the controller gain k.

```
import sys

r = float(sys.argv[1]) # Reference or "setpoint"
k = float(sys.argv[2]) # Controller gain

u = 0          # "Previous" output
for _ in range( 200 ):
    y = u          # One-step delay: previous output

    e = r - y    # Tracking error
    u = k*e        # Controller output

    print r, e, 0, u, y
```

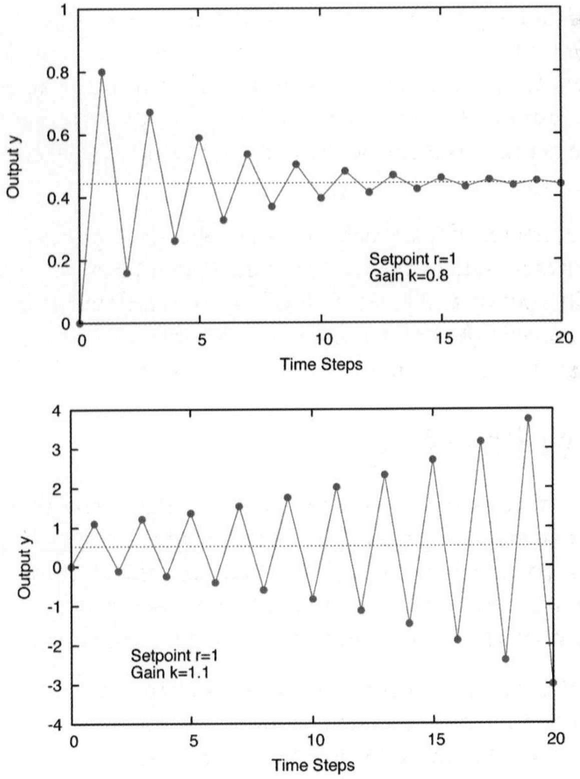

*Figure 3-4. Time evolution of the system $y_{t+1} = k(r - y_t)$ for two differ-
ent values of k.*

Although the setpoint is constant, the process output oscillates. Moreover, for values of the controller gain k greater than 1, the amplitude of the oscillation grows without bounds: The system diverges.

If you look closely, you will also find that the value to which the system converges (if it does converge) is *not* equal to the desired setpoint. (You might want to base control actions on the cumulative error, as in Chapter 2. Does this improve the behavior?)

CHAPTER 4

Controllers

The purpose of a *controller* is to produce a signal that is suitable as input to the controlled plant or process. Controllers occur in both open-loop configurations (Figure 4-1) and closed-loop configurations (Figure 4-2).

Figure 4-1. Open-loop control configuration consisting of controller K and the controlled system H. Boxes indicate systems, and control signals flow in the direction of the arrows.

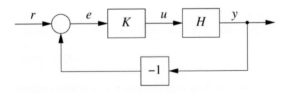

Figure 4-2. Closed-loop control configuration consisting of controller K and the controlled system H. The open circle forms the sum of all incoming signals.

The need for a controller—simply to perform numerical transformations—becomes apparent if we consider some examples. In the case of the heated vessel, the controller input will be a temperature value, but the input to the heating element itself will be a voltage, so if nothing else we need to transform units and numerical values. In the case of the read-through cache, the controller input is a hit rate and so, by construction, a number between 0 and 1 in magnitude. In contrast, the size of the cache is always positive and possibly quite large (hundreds or thousands of elements). Again, there is (at least) a need to perform a transformation of the numerical values.

Beyond the common need to transform numerical values, open-loop (feedforward) and closed-loop (feedback) configurations put different demands on a controller. In the open-loop case, the controller must be relatively "smart" in order to compensate for the complexities of the plant and its environment. By contrast, controllers used in closed loops can be extremely *simple* because of the self-correcting effect of the feedback path. *Feedback systems trade increased complexity in overall loop architecture for a simpler controller.*

Although any component that transforms an input to an output can be used as a controller in a feedback loop, only two types of controller are encountered frequently: the on/off controller and the three-term (or PID) controller. Both can be used in a feedback loop to transform their input (namely the tracking error) into a signal that is suitable as input to the controlled plant.

Block Diagrams

The structure of a control loop is easily visualized in a *block diagram* (see Figure 4-1 and Figure 4-2). Block diagrams consist of only three elements as follows.

Boxes:
> Boxes represent systems, such as controllers, filters, and controlled processes.

Arrows:
> Arrows show the flow and direction of control signals.

Circles:
> Open circles indicate summers: all control signals arriving at an open circle are summed to form the element's output.

The takeoff point for a signal is sometimes indicated by a small filled dot (as for the y signal in the extreme right of the closed-loop diagram in Figure 4-2). The box labeled −1 has the effect of changing the sign of its input.

There exists a collection of rules (*block-diagram algebra*) for manipulating block diagrams. These rules make it possible to transform and simplify diagrams purely graphically—that is, without recourse to analytic expressions—while maintaining their correct logical meaning (see Chapter 21).

On/Off Control

The simplest type of controller consists of nothing more than a plain on/off switch: whenever the tracking error is positive (that is, when the plant output is *below* the desired setpoint), the plant is being "turned on full"; whenever the tracking error is negative, the plant is being turned off. Such controllers are sometimes known as "*bang-bang controllers.*" See Figure 4-3.

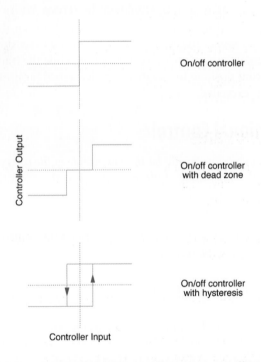

Figure 4-3. Three forms of on/off controller: plain on/off control (top), on/off control with dead zone (center), and on/off control with hysteresis (bottom).

Although deceptively simple, this control strategy has serious draw-backs that make it unsuitable for many practical applications. The main problem is that the system never settles down to a steady state; instead, it oscillates *constantly* and *rapidly* between its two extreme states. Think of a car with cruise control (which is a feedback system designed to maintain a constant speed) operating this way: instead of maintaining a steady 65 mph, an on/off cruise control would open up the throttle full whenever the speed falls even a fraction below the reference speed, only to return the engine to idle as soon as the speed exceeds the setpoint again. Such operation would be hard on the en-gine, the transmission, and the suspension—not to mention the pas-sengers!

We can improve on/off controllers by augmenting them with a strategy to inhibit such rapid control oscillation. This can be done either by introducing a dead zone or by employing hysteresis. With a dead zone, the controller will not send a signal to the plant unless the tracking error exceeds some threshold value. When using hysteresis, the con-troller maintains the same corrective action while the tracking error switches from positive to negative (or vice versa), again until some threshold is exceeded.

Proportional Control

It is a major step forward *to let the magnitude of the corrective action depend on the magnitude of the error*. This has the effect that a small error will lead to only a small adjustment, whereas a larger error will result in a greater corrective action.

The simplest way to achieve this effect is to let the controller output be *proportional* to the tracking error:

$$u_p(t) = k_p e(t) \qquad k_p > 0 \text{ constant}$$

where k_p, the *controller gain*, is a positive constant.

Why Proportional Control Is Not Enough

Strictly proportional controllers respond to tracking errors—in par-ticular, to *changes* in the tracking error—with a corrective action in the correct direction; but in general they are insufficient to eliminate tracking errors in the steady state entirely. When using a strictly proportional controller, the system output y will always be *less* than the desired setpoint value r, a phenomenon known as *proportional droop*.

The reason is that a proportional controller, by construction, can produce a nonzero output only if it receives a nonzero input. If the tracking error vanishes, then the proportional controller will no longer produce an output signal. But most systems we wish to control will require a nonzero input in the steady state. The consequence is that some residual error will persist if we rely on purely proportional control.

Proportional droop can be reduced by increasing the controller gain k, but increasing k by too much may lead to instability in the plant. Hence, a different method needs to be found if we want to eliminate steady-state errors. Of course, one can intentionally adjust the setpoint value to be higher than what is actually desired—so that, with the effect of proportional droop, the process output settles on the proper value (a process known as "manual reset"). But it turns out that this manual process is not even necessary because there is a controller design that can eliminate steady-state errors automatically. This brings us to the topic of integral control.

Integral Control

The answer to proportional droop—and, more generally, to (possibly small) steady-state errors—is to base the control strategy on the total accumulated error. The effect of a proportional controller is based on the *momentary* tracking error only. If this tracking error is small, then the proportional controller loses its effectiveness (since the resulting corrective actions will also be small). One way to "amplify" such small steady-state errors is to keep adding them up: over time, the accumulated value will provide a significant control signal. On the other hand, if the tracking error is zero, then the accumulated value will also be zero. This is the idea behind integral control.

The output of an integral controller is proportional to the *integral* of the tracking error over time:

$$u_i(t) = k_i \int_0^t e(\tau)\, d\tau \qquad k_i > 0 \text{ constant}$$

Bear in mind that an integral is simply a generalization of taking a sum. In a computer implementation, where time progresses in discrete steps, it becomes a sum again. An integral controller is straightforward to implement as a cumulative sum of the error values. This approach lends itself to a convenient "recursive" updating scheme:

$$E_t = \delta t \cdot e_t + E_{t-1}$$

$$u_{i,t} = k_i E_t$$

where E_t is the cumulative error at time step t, k_i is the *integral gain*, and $u_{i,t}$ is the output of the integral controller at time t.

This discrete updating scheme assumes that control actions are performed periodically. The factor δt is the length of time between successive control actions, expressed in the units in which we measure time. (If we measure time in seconds and make 100 control actions per second, then $\delta t = 0.01$; if we measure time in days and make one update per day, then $\delta t = 1$.) Of course, δt could be absorbed into the controller gain k_i, but this would imply that if we change the update frequency (for example, by switching to two updates per day instead of one), then the controller gains also would have to change! It is better to keep things separate: δt encapsulates the time interval between successive updates, and k_i independently controls the contribution of the integral term to the controller output.

It is common to use both a proportional and an integral controller in parallel (see Figure 4-4). In fact, this particular arrangement—also known as a PI (proportional-integral) controller—is the variant most frequently used in applications.

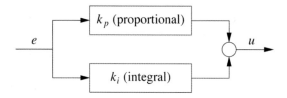

Figure 4-4. Block diagram of the two-term (or PI) controller. The parameters k_p and k_i are the controller gains.

Integral Control Changes the Dynamics

The output of an integral controller depends not only on the momentary value of the error but also on the integral (or the sum) of the observed tracking errors since the beginning of time. This dependence on past values implies that an integral controller possesses nontrivial dynamics (as discussed in Chapter 3), which may change the qualitative behavior of the entire loop.

In particular, an integral controller may introduce oscillations into a loop even if the controlled system itself is not capable of oscillations. If a positive tracking error persists, then the integral term in the controller will increase; the result will be a positive input to the plant that will persist even after the tracking error is eliminated. As a consequence, the plant output will overshoot and the tracking error will become negative. In turn, the tracking error reduces the value of the integral term.

Depending on the values chosen for the controller gains k_p and k_i, these oscillations may decay more or less quickly. Controller tuning is to the process of finding values for these parameters that lead to the most acceptable dynamic behavior of the closed-loop system (also see Chapter 9).

Integral Control Can Generate a Constant Offset

One of the usual assumptions of control theory is that the relationship between input and output of the controlled plant is *linear*: $y = H\, u$. This implies that, in the steady state, there can be no nonzero output unless there is a nonzero input. In a feedback loop we try to minimize the tracking error, but we also use the tracking error as input to the controller and therefore to the plant. So how can we maintain a nonzero output when the error has been eliminated?

We can't if we use a strictly proportional controller. That's what "proportional droop" is all about: under proportional control, the system needs to maintain some residual, nonzero tracking error in order to produce a nonzero output. But we *can* drive the tracking error all the way down to zero and still maintain a nonzero plant output—provided that we include an integral term in the controller.

Here is how it works: consider (again) the heated pot on the stove. Assume that the actual temperature (that is, the system output y) is below the desired value and hence there is a nonzero, positive tracking error $e = r - y$. The proportional term multiplies this error by the gain to produce its control signal ($u_p = k_p e$); the integral term adds the error to its internal sum of errors ($E = E + e$) and reports back its value ($u_i = k_i E$). In response to the combined output of the controller, more heat is supplied to the heated pot. Assume further that this additional heat succeeds in raising the temperature in the pot to the desired value. Now the tracking error is zero ($e = 0$) and therefore the control signal due to the *proportional* term is also zero ($u_p = k_p \cdot 0 = 0$). However, the internal sum of tracking errors maintained by the *integral* term has not changed. (It was "increased" by the current value of the tracking error, which is zero: $E = E + 0 = E$.) Hence the integral term continues to produce a nonzero control signal ($u_i = k_i E$), which leads to a nonzero process output. (In regards to the heated pot, the integral term ensures that a certain amount of heat continues to be supplied to the pot and thereby maintains the desired, elevated temperature in the vessel.)

Derivative Control

Finally, we can also include a derivative term in the controller. Whereas an integral term keeps track of the past, a derivative controller tries to anticipate the future. The derivative is the rate of change of some quantity. So if the derivative of the tracking error is positive, we know that the tracking error is currently *growing* (and vice versa). Hence we may want to apply a corrective action immediately, even if the value of the error is still small, in order to counteract the error growth—that is, before the tracking error has a chance to become large.

Therefore, we make the output of the derivative controller proportional to the derivative of the tracking error:

$$u_d(t) = k_d \frac{de(t)}{dt} \qquad k_d > 0 \text{ constant}$$

In a discrete-time computer implementation, we can approximate the derivative of e by the amount e has changed since the previous time step. A derivative controller can thus be implemented as

$$u_{d,t} = k_d \frac{e_t - e_{t-1}}{\delta t}$$

where δt is the time interval between successive updates (as for the integral term).

Like the integral controller, the derivative controller depends on past values and therefore introduces its own, nontrivial dynamics into the system.

Problems with Derivative Control

Whereas integral controllers are extremely "benevolent" and so are often used together with proportional controllers, the same cannot be said about derivative control. The problem is the potential presence of high-frequency noise in the controller input.

The noise contribution will fluctuate around zero and will thus tend to cancel itself out in an integral controller. (In other words, the integrator has a smoothing effect.) However, if we take the derivative of a signal polluted by noise, then the derivative will enhance the effect of the noise. For this reason, it will often be necessary to *smooth* the signal. This adds complexity (and additional nontrivial dynamics) to the controller, but it also runs the risk of defeating the purpose of having derivative control in the first place: if we oversmooth the signal, we will eliminate exactly the variations in the signal that the derivative controller was supposed to pick up!

Another problem with derivative control is the effect of sudden setpoint changes. A sudden change in setpoint will lead to a very large momentary spike in the output of the derivative controller, which will be sent to the plant—an effect known as *derivative kick*. (In Chapter 10 we will discuss ways to avoid the derivative kick.)

Whereas proportional control is central to feedback systems, and integral control is required in order to eliminate steady-state errors, it should come as no surprise that derivative control is less widely used in practice. Studies show that as many as 95 percent of all controllers used in certain application areas are of the proportional-integral (PI) type.

The Three-Term or PID Controller

A controller including all three components (proportional, integral, and derivative) is known as a *three-term* or *PID controller* (see Figure 4-5). Its output is a combination of its three components:

$$u_{\text{PID}}(t) = u_p(t) + u_i(t) + u_d(t)$$

$$= k_p e(t) + k_i \int_0^t e(\tau)\,d\tau + k_d \frac{de(t)}{dt}$$

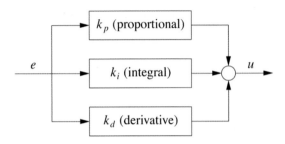

Figure 4-5. In addition to the proportional and integral terms, the three-term (or PID) controller also contains a derivative term with gain k_d.

This is the form most convenient for theoretical work. In application-oriented contexts, an alternative form is often used that factors out an overall gain factor:

$$u_{\text{PID}}(t) = k\left[e(t) + \frac{1}{T_i} \int_0^t e(\tau)\,d\tau + T_d \frac{de(t)}{dt} \right]$$

The new parameters T_i and T_d both have the dimension of *time*. The two formulas for $u_{\text{PID}}(t)$ are equivalent, and their parameters are related:

$$k = k_p \qquad T_i = \frac{k_p}{k_i} \qquad T_d = \frac{k_d}{k_p}$$

Of course, the *numerical values* of the parameters are different! When comparing values for controller parameters, one must not forget to establish which of the two forms they refer to.

Convention

In this book, controller gains are always nonnegative. (That is, they are zero or greater.)

We'll pick up the study of PID controllers again in Chapter 22, by which time we will have acquired a larger set of theoretical tools.

Code to Play With

The discrete-time updating schemes for the integral and derivative terms lend themselves to straightforward computer implementations. The following class implements a three-term controller:

```
class PidController:
    def __init__( self, kp, ki, kd=0 ):
        self.kp, self.ki, self.kd = kp, ki, kd
        self.i = 0
        self.d = 0
        self.prev = 0

    def work( self, e ):
        self.i += DT*e
        self.d = ( e - self.prev )/DT
        self.prev = e

        return self.kp*e + self.ki*self.i + self.kd*self.d
```

Here the factor DT represents the step length δt, which measures the interval between successive control actions, expressed in the units in which time is measured.

This controller implementation is part of a simulation framework that can be used to explore control problems. In Chapter 12, we will discuss this framework in more detail and also introduce a better controller implementation that avoids some deficiencies of the straightforward version presented here.

CHAPTER 5

Identifying Input and Output Signals

Initially, it can be difficult to see how feedback methods can be applied to situations other than the "classical" application areas treated in textbooks on control theory. The way control engineering decomposes real-world systems into abstractions often does not easily align with the way those systems appear to others. In this chapter, I want to step through a handful of examples and show how they could be approached from a control-theoretic point of view.

Control Input and Output

The essential abstraction in any control problem is the *plant* or *process*: the system that is to be controlled. From a controls perspective, a plant or process is a black box that transforms an input to an output. It is usually not difficult to recognize the plant itself, but identifying what to use as control *input* and *output* can be challenging.

It is essential to realize that the terms "input" and "output" here are used only in relation to the *control* problem and may be quite different from the *functional* inputs and outputs of the controlled system. Specifically:

- The *input* or *control input* is a quantity that we *can* adjust directly. By adjusting the input, we hope to influence the output in a favorable way.

- The *output* or *process output* is the quantity we *want* to control: we want the output to track the reference value (the setpoint).

These two observations should help to identify the quantities to use as control signals, either as input or as output. Just ask yourself:

- What quantity can we influence directly?
- What quantity do we (ultimately) want to influence?

In books on control theory, the output is often referred to as the "process variable" (PV) and the input as the "control variable" or "manipulated variable" (MV). Both taken together define the system's "interface" (in a software-engineering sense of the word).

There may be situations where both of these quantities are identical. In this case, you are done and you can stop reading. But often they will not be the same—and then you have a control problem.

Directionality of the Input/Output Relation

The basic idea of feedback control is to compare the actual plant output y to the reference value r and then to apply a corrective action that will reduce the tracking error $e = r - y$. In order to do so, we must know in which *direction* to apply the correction. Let's say that the output y is smaller than what it should be ($y < r$), so that the tracking error is positive. Obviously, we want to increase the process output—but does this mean we need to *increase* the plant input u or *decrease* it?

The answer depends on the *directionality* of the input/output relation for the controlled system. It is usually assumed that increasing the control input will increase the control output:

- *Increasing* the power supplied to a heating element (the input) will *increase* the temperature of the heated room or vessel (the output).
- *Increasing* the number of servers in a data center will *increase* the number of requests handled per hour.

However, the opposite also occurs:

- *Increasing* the power supplied to a cooling unit (the input) will *decrease* the temperature in the cooled room or vessel (the output).

- *Increasing* the number of servers in a data center will *reduce* the average response time for server requests.

The directionality depends on the *specific* choice of input and output signals, not on the overall plant or process, as is demonstrated by the data center example. Depending on the particular choice of output signal, the same system can exhibit either form of directionality.

We need to take the input/output directionality into account when designing a control loop to ensure that corrective actions are applied in the appropriate direction. A standard loop (Figure 5-1) is suitable for the "normal" case, where an increase in control input results in an increase in control output. For the "inverse" case (where an increase in control input leads to a *decrease* in output), we can use a loop as in Figure 5-2. In this loop, the tracking error e is multiplied by -1 before being passed to the controller.[1]

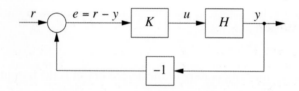

Figure 5-1. *The standard loop arrangement for systems where an increase in plant input u leads to an increase in plant output y.*

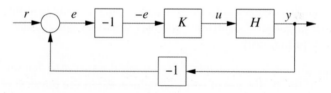

Figure 5-2. *A loop arrangement suitable for systems with an inverted input/output relation, where an increase in plant input u leads to a decrease in plant output y.*

1. Of course, it is possible to absorb this step into the controller by using negative gains. However, I find it convenient to follow the convention that controller gains are always nonnegative and also to make the deviation from the standard loop architecture explicit by introducing the additional inverter element. In this way, the controller is always a "normal" element, for which an increase in input leads to an increase in output.

Examples

Let's consider a few examples and discuss the available options for both control inputs and outputs as well as their advantages and disadvantages. When evaluating each situation, we are primarily concerned with two properties: which signals are *available* and which signals will show the *speediest* response. (In Part III, we will study many of these examples in more depth.)

Thermal Control 1: Heating

Situation

A room is to be kept at a comfortable temperature, or a vessel containing some material is to be kept at a specific temperature.

Input

The input is the amount of heat applied; this may be the dial setting on the stove, the voltage applied to the heating element, or the flow of heating oil to the furnace.

Output

In any case, the temperature of the heated object is the output of the process.

Commentary

This example is a conventional control problem, where the control strategy—including the choice of input and output signals—is pretty clear. Notice that the control strategy can vary: central heating often has only an on/off controller, whereas stoves let you regulate their power continuously.

Although simple in principle, this example does serve to demonstrate some of the challenges of control engineering in the physical world. We said that the output of the plant is "the temperature" and that its input is "the heat supplied." Both are physical quantities that are not easy to handle directly. Just imagine having to develop a physical device to use as a PID-controller that takes "a temperature" as input and produces a proportional amount of "heat" as output! For this reason, most control loops make use of electronic control signals. But this requires the introduction of additional elements into the control loop that can turn electric signals into physical action and observed quantities into electric signals (see Figure 5-3, top).

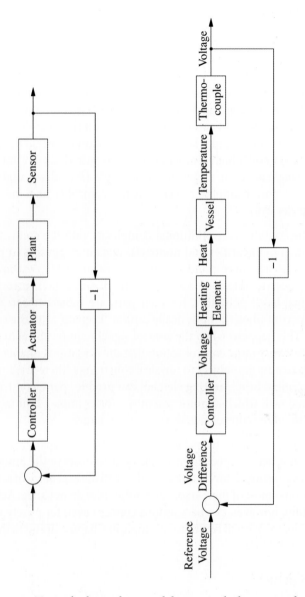

Figure 5-3. Typical physical control loops, including transducers to convert between physical quantities and control signals (which are usually electrical signals). Top: Schematic. Bottom: A control loop configuration for the heated vessel example, showing the observed quantities and control signals in some detail.

Elements that convert back and forth between control signals and physical quantities are generally known as *transducers*. A transducer that measures a physical quantity and turns it into a control signal is called a *sensor*, and a transducer that transforms a control signal into a physical control action is called an *actuator*. In the case of the heated vessel, the sensor might be a thermocouple (transforming a measured temperature into a voltage), and the actuator would be a heating element, transforming an applied voltage into heat (Figure 5-3, bottom). If the vessel to be heated is large (such as a reaction vessel in the chemical industry) or if high temperatures are required (as in a furnace), then a single actuator might not be enough; instead we might find additional power amplifiers between the controller and the actual heating element.

Transducers introduce additional complexity into the control loop, including nonlinearities and nontrivial dynamics. Sensors in particular may be slow to respond and may also be subject to *measurement noise*. (The latter might need to be smoothed using a filter, introducing additional lags.) Actuators, for their part, are subject to *saturation*, which occurs when they physically cannot "keep up" with the control signal. This happens when the control signal requires the actuator to produce a very large control action that the actuator is unable to deliver. Actuators are subject to physical constraints (there are limits on the amount of heat a heating element can produce per second and on the speed with which a motor can move, for example), limiting their ability to follow control signals that are too large.

In computer systems, transducers rarely appear as separate elements (we do not need a special sensor element to observe the amount of memory consumed, for instance), and—with the exception of actuator saturation—most of their associated difficulties do not arise. Actuator saturation, however, can be a serious concern even for purely virtual control loops. We will return to this topic in Chapter 10 and in the case studies in Part III.

Item Cache

Situation

Consider an item cache—for instance, a web server cache. If a request is made to the system, the system first looks in the cache for the requested item and returns it to the user if the item is found. Only if the item is not found will the item be retrieved from persistent storage.

The cache can hold a finite number of items, so when an item is fetched from persistent storage, it replaces the oldest item currently in the cache. (We will discuss this system in detail as a case study in Chapter 13.)

Input

The quantity that can be controlled directly is the number of items in the cache.

Output

The quantity that we want to influence is the resulting "hit" or success rate. We desire that some fraction (such as 90 percent) of user requests can be completed without having to access the persistent storage mechanism.

Commentary

Observe that the quantities identified as "input" and "output" *from a control perspective* have nothing to do with the incoming user requests or the flow of items into and out of the cache. This is a good example of how the control inputs and outputs can be quite different from the *functional* inputs and outputs of a system—it is important not to confuse the two!

Also, the definition of the output signal is still rather vague. What does "90 percent hit rate" mean in practice? Ultimately, each hit either succeeds or fails: the outcome is binary. Apparently we need to average the results of the last n hits to arrive at a hit rate, which leads us to ask: how large should n be? If n is large, then the outcome will be less noisy but the *memory* of the process will be longer, so it will respond more slowly to changes (see Chapter 3). Selecting a specific value for n therefore involves a typical engineering trade-off.

We also need to define how the average is to be taken. Do we simply calculate a straight average over the last n requests? Or should we rather weight more recent requests more heavily than older requests? For practical implementations, it may be convenient to employ an exponential smoothing method (a recursive digital filter), where the value of the hit rate s_t at time t is calculated as a mixture of the outcome of the most recent user request σ_t and the previous value of the hit rate:

$$s_t = \alpha \sigma_t + (1 - \alpha) s_{t-1} \quad 0 < \alpha < 1$$

Here, σ_t is either 0 or 1, depending on the outcome of the most recent cache request. The smaller α is, the smoother the signal will be. (In fact, we can choose whether to consider the filter as part of the *system* itself or to treat it as a separate component. In the former case, the system output will be the smoothed hit rate; in the latter case, the system output will be the string of 0's and 1's corresponding to the outcome of the most recent request.)

Finally, we should discuss the notion of "time" t. From the preceding remarks it is clear that the system will not produce a new and different value of its output unless a user request is made. We must decide when control inputs (that is, changes to the size of the buffer) can be made: only synchronously with user requests, or asynchronously at any time that we desire (for instance, periodically once per second)? In the asynchronous case, the main "worker thread" of the cache (the one that handles user requests) will be separate from the "control thread" that handles control inputs. This is yet another reminder that control flow and functional flow are quite separate things.

Server Scaling

Situation

Imagine a central server performing some task. It could be a web server serving user requests, or a compute server performing CPU-intensive jobs, or even a DB server. In any case, we can control the number of active "worker" instances (the number of threads in the case of the web server, or the number of CPUs for the compute server, and so on). Incoming requests are assigned to the next available worker instance; it will take the worker some (random) amount of time to complete each request. Requests that cannot be served immediately are submitted to a queue. There is no guarantee that the throughput of the server will scale linearly with the number of active worker instances! (More specific systems fitting this general description will be discussed in detail in Chapter 15 and Chapter 16.)

Input

We can select the number of active worker instances directly.

Output

Ultimately, we want to make sure tasks are flowing through the system without being held up. However, we have a wide selection of metrics that can be used for that purpose. These include:

- Number of requests queued (the queue length)
- Net change in the length of the queue (over some time interval)
- Average age of requests in queue
- Maximum age of requests in queue (age of oldest request)
- Total age (since its arrival) of the last *completed* request
- Average age (since arrival) of the last k completed requests
- Requests completed in the last T seconds
- Fraction of idle time (over the last T seconds) across all active worker instances

Which of these quantities to use will be one of the design choices.

Commentary

In this example, we observe again that the functional flow of requests into the controlled system has nothing to do with the input and output signals that we use for control purposes. Moreover, and also as in the previous example, we find that the control input is easy enough to find, yet we have considerable freedom in the definition of the process output.

From a business domain perspective, the maximum age among all the requests in the queue seems like a good metric because it represents a quantity that is immediately relevant: the worst-case waiting time. However, because it depends on a *single* element only, this metric will be noisier than the *average* waiting time across all elements currently in the queue. On the other hand, the average waiting time responds more slowly to changes in the environment because the effect of any change is "averaged out" over the number of items in the queue. This is not desirable: we want signals that make any change visible quickly so that we can respond to them without delay.

Any form of smoothing or averaging operation slows signals down, but a poor choice of raw signal can also result in delayed visibility. The age of the last completed request has this property: if we use this quantity as a control signal, then we will not learn about the growth of the queue until every item has actually propagated through the queue to a worker instance! With this choice of monitored quantity, we are in the undesirable position that—by the time we learn what's going on—it's already too late to do something about it. For this reason, we want to begin monitoring items as soon as possible, that is, when they enter the queue and not when they leave it.

Although the maximum or average age of items in the queue may be a desirable control signal from a business domain perspective, it may not be available on technical grounds. We may not know the time stamp for each item's arrival simply because this information is not being recorded! In this case, we may have to fall back on using the number of items currently in the queue. This quantity has the advantage that it is both easy to obtain and responds quickly to changes, but it is less directly related to the property that we really want to control. (The queue length actually does not matter much provided that items are being processed at a rapid rate.) In the end, the net change in the length of the queue may seem like a good compromise: it responds quickly, does not depend on a single item, and is relevant from a domain perspective. Too bad it cannot be observed directly! We must calculate it as the difference between the previous and the current queue length—a procedure that still requires us to fix the interval at which we observe the queue length.

All the quantities discussed so far tend to *decrease* as the input quantity (namely the number of active worker instances) is increased, so we will need to use a loop structure suitable for this kind of input/output directionality. In contrast, the last two items in the list of possible output signals (requests completion rate and amount of idle time) tend to increase with the number of server instances available.

Controlling Supply and Demand by Dynamic Pricing

Situation

Consider a merchant selling some arbitrary product. The merchant has the goal of selling a certain number of units every day; the merchant's primary control mechanism is the item price, which can be adjusted on a daily basis. (An application of this situation is discussed as a case study in Chapter 14.)

Input

The price per unit.

Output

The number of units sold.

Commentary

In this example, the choice of input and output quantities is obvious; but what exactly constitutes the "plant" and its dynamics merits some discussion.

The input and output for this problem (namely, price and units sold) are related through what economists call the *demand curve* (although they tend to interchange the respective roles of price and demand). Typical shapes of the demand curve—using our identification of independent and dependent variables—are shown in Figure 5-4. (Note that these curves, too, have the property that increasing the input leads to a decrease in output, which makes the use of the "inverted" loop structure necessary.)

If the merchant *knew* this curve, then there would be no need for a feedback system to control demand: the merchant could simply pick the exact price that would result in the sale of the desired number of units. However, in general the merchant does *not* know the demand curve—hence the need for a system that automatically applies corrective actions as needed.

So then where is the "memory" that, according to the discussion in Chapter 3, is the hallmark of systems with nontrivial dynamics? At first, it might appear as if there is no memory: at the beginning of each day, the vendor can fix a new price that becomes effective immediately. Nevertheless, the memory is there—in this case, it is quite literally in the vendor's head! The feedback mechanism will not work if the merchant *randomly* chooses a new price every day. Instead, for the feedback loop to be closed, the price the merchant quotes tomorrow must be based on the number of units sold today, which in turn is a consequence of today's price. The feedback controller in this case does not so much produce a new price as it produces an *update* to the current price.

So far we have assumed that the demand curve itself does not change over time—or, at least, that it changes sufficiently slowly that these changes can be ignored for day-to-day price finding. Furthermore, we assumed that the price–demand relationship is deterministic, without random variations. That's not likely to be the case in practice (demand typically fluctuates), but this does not pose a fundamental challenge. The demand could be averaged over a few days to obtain a smooth control signal, even though this operation will introduce some inevitable delays.

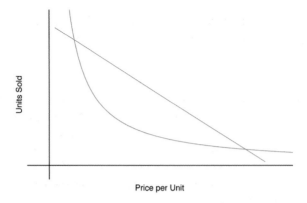

Figure 5-4. Two typical shapes of the demand curve, which relates the number of units sold to their price.

We also have not mentioned how the merchant chooses the number of units to sell every day. Couldn't the merchant make more money by selling more units at a lower price? That's an interesting question, but it is *extremely important* to remember that feedback control does not provide any help in answering it! Feedback control is a mechanism to track a reference signal—not more, not less. In particular, feedback control makes no statement, which value for the setpoint is "optimal." That question must be answered separately; once it has been answered, feedback can be used to execute on this plan.

Finally, keep in mind that the terms "merchant," "price," and "demand" in this example are largely metaphorical. Similar considerations apply in other situations, as long as there is a long sequence of fundamentally similar transactions that are being completed over time. A consumer, repeatedly buying from a supplier, fits the same model. Note also that the "price" or "cost" need not be measured in monetary units. For example, the task server example discussed previously can be expressed in these terms—provided the server is able to report on the "effort" expended in a way that can be used as a control signal.

Thermal Control 2: Cooling

Situation

Many computer systems require active cooling (using fans) to keep components at acceptable operating temperatures. (See Chapter 17 for an in-depth discussion of this example.)

Input

The input is the fan speed or the voltage applied to the fan—the details will depend on the interface provided by the system.

Output

It may appear as if the process we want to control is the CPU and so its temperature should be the output, but this is not quite right. The process we want to control is the *cooling* of the CPU by the fan; therefore the proper way to measure this process is the *reduction* in temperature achieved. Remember that if the process is off (that is, if the fan is not running) the output should be zero, and that the process output should increase in line with the input. (Yet another example of an "inverted" output signal.)

Commentary

This example is interesting, because the *physics* is exactly the same as in the heating example that opened this chapter, yet many details that are relevant for a control application are different. We have already seen that we need to be careful with the choice of output signal. Another possible misidentification concerns the process dynamics.

The temperature of a computer component has its own dynamics, which are the same as that of any other heated element: once switched on, the temperature will increase and eventually reach a thermal equilibrium in which the component gives off as much heat as is being supplied externally. But that is not the dynamic we want to control! Instead, the dynamic that we care about from a control perspective is how quickly the temperature of the chip *drops* once the fans are turned on. The "off" state of this process is a chip in thermal equilibrium, not a chip that is switched off.

But this poses an interesting operational problem: a chip in thermal equilibrium is "fried" and does not function anymore (otherwise, we wouldn't need active cooling to begin with). Therefore, the baseline is a chip operating at its maximum permissible temperature, with just enough cooling being supplied to keep it there. As the fan speed increases, we can see how much and how quickly the temperature drops and thereby observe the dynamics of the actual control process.

Criteria for Selecting Control Signals

There will frequently be more than one candidate quantity that can be used as control input (or output), and we are free to choose from among them. This choice amounts to an engineering decision, and we obviously want to use those signals that have the most favorable characteristics. In this section, we discuss some criteria by which to evaluate different possibilities.

For Control Inputs

Any quantity that is considered a candidate for being a control input should be evaluated according to the following criteria.

Availability:
> Only quantities that we can influence directly and immediately are suitable as control inputs.

Responsiveness:
> The system should respond quickly to a change in its input in order to obtain good dynamic performance and accurate tracking in the presence of change. Try to avoid inputs whose effect is subject to latency or delays.

Granularity:
> It is desirable to be able to adjust the control input in small increments to achieve accurate tracking. The PID controller in particular requires a system that is capable of responding to the continuous output this controller produces. That is not always possible—the number of server instances in a server farm, for instance, can be changed only in integer increments. If the "right" number of servers is not a whole integer, then the system will not reach a steady state under PID control. If a system's output can be adjusted only in large, fixed increments, then it may be necessary to modify the controller or introduce special-purpose actuators to obtain satisfactory control behavior. (The case studies in Chapter 15, Chapter 16, and Chapter 18 discuss some possibilities.)

Directionality:
> Does increasing the input result in an increase or a decrease of the chosen output? If an increase in input leads to a reduction in output, then an "inverted" loop must be used.

For Control Outputs

To evaluate the suitability of a quantity as a control output, we should consider the following criteria.

Availability:
> The quantity must be observable—accurately, reliably, and quickly—without gaps in coverage and without delays.

Relevance:
> The output signal should be a good measure for the behavior that we want to control. This is a nonissue if the output itself is the quantity to be controlled (as in the heating example earlier). But if we are interested in measuring the system's overall "quality of service," then a variety of metrics can be used as proxy for this abstract idea, and we should be careful to choose the one that is most informative with respect to the intended purpose. (The task server example in this chapter was of this kind.)

Responsiveness:
> The output metric should reveal changes in the system's state or behavior quickly. This means avoiding lags and delays. Lags typically occur when the output metric is calculated as an "average" over a set of values, whereas delays occur when some quantity needs to "propagate" through the system in order to become observable. (See Chapter 3.)

Smoothness:
> In a closed-loop arrangement, the output is part of the control input. Disturbances (such as discontinuities or noise) in the output will therefore result in sudden control actions— something we usually want to avoid. For this reason, it is desirable to choose an output signal that is already relatively smooth and does not need to be filtered. But watch out: signals that are naturally smooth may, in fact, be the result of an implicit filtering or averaging process (inside the controlled system) and therefore subject to lags.

With output signals especially, we must make trade-offs between the various desirable properties on a case-by-case basis.

A Note on Multidimensional Systems

You may have noticed that all examples used only *scalar* input and output signals. In each example, we used a single input control parameter to control a single output metric. This raises the question of whether it is possible to construct control loops using multiple inputs and outputs simultaneously.

This is certainly possible, but it is much more difficult, because the various input and output signals will typically not be *independent*. That is, changing *one* of the input signals will usually lead to changes in *several* of the output signals. This prevents naive application of the feedback principle (constantly compare the output to the setpoint and then apply a corrective action that counteracts the deviation from the setpoint—see Chapter 2), because we won't be able to determine the proper "direction" for the corrective action. For scalar control signals, it is relatively easy to determine whether we need to increase or decrease the control input in order to reduce the output (because there are only these two possibilities), but for a system with several inputs, things are no longer so simple. The number of different control input combinations increases rapidly with the dimension of the control signal, and determining the input/output relationships (including the interactions between different input signals) from experiments alone will usually be impractical. Hence control situations involving multiple input and output signals pretty much require a good *theoretical* process model.

Even with a good model, controlling several outputs simultaneously is a difficult problem. A general approach is to try and *decouple* the various signals in order to reduce the multidimensional control problem to a set of scalar ones. Is one of the control inputs clearly dominant? If so, then we can try basing the entire control strategy on that signal alone. Another possibility is to try decoupling multiple signals into separate loops. If the system will respond to one of the signals much faster than to another, then we can often treat these two signals as independent, and there are special loop arrangements for this situation ("cascaded control"; see Chapter 11).

CHAPTER 6

Review and Outlook

Before proceeding to practical matters that are directly relevant to applications and implementations, let's summarize the conceptual foundations of feedback control as we have developed them so far.

The Feedback Idea

The idea behind feedback control is simple:

> *Constantly compare the actual output to the setpoint; then apply a correction in the correct direction and of approximately the correct size.*

The comparison and corrections are performed at runtime. Precisely because feedback relies on runtime observations and adjustments, feedback control is capable of responding to unanticipated disturbances.

Iteration

Feedback is an *iterative* scheme. That we keep monitoring the output and applying corrections is what makes feedback control feasible. Instead of having to get it "right" in a single step, we need only make things "better" because there is always another chance to fix any outstanding errors.

As a side effect, the ongoing iteration will also make the system robust to change.

Process Knowledge

One benefit of the feedback concept is that it does not require detailed knowledge about the controlled system and its behavior. Only two bits of information are required:

- We *must* be able to identify the correct direction for the application of a corrective action. (In other words, we must know whether increasing the input will end up increasing or decreasing the output. This amount of process knowledge is indispensable.)

- In general, we want to apply the *largest* possible correction that will not make the system unstable, in order to achieve the quickest possible reduction and elimination of the tracking error. This implies that we must be able to estimate the typical size or scale of the system's response to an input change.)

Although feedback control does not require detailed knowledge about the controlled process, we must have at least enough information to answer the two preceding items in order to apply feedback control successfully. (Incidentally, it is this requirement that makes multidimensional control so hard: obtaining even these insights is extremely difficult if there is more than one control signal involved.)

Avoiding Instability

A system exhibits unstable behavior if it permanently oscillates between over- and undercompensation, without converging to a steady state. In extreme cases of instability, the amplitude of the oscillations increases over time (until the system is destroyed). Instability is usually the result of control actions that are too large.

The "theory" primarily tries to determine how large control actions can be for a given system. The answer depends on the static "scale" of the system's input/output relation in the steady state *and* on the dynamic response of the system to an input change. Any form of lag or delay typically has the effect of reducing the magnitude of the corrective action that can be applied in any given moment.

The Setpoint

Feedback control has the effect of reducing the tracking error, which is the difference between the reference value (or *setpoint*) and the actual process output.

A necessary ingredient for feedback control is the existence of such a setpoint. We must have a notion of a desired value for the tracked metric. If we cannot identify a setpoint and cannot formulate a specific value (or, at least, a range of values) to track, then feedback control is *not applicable*.

Control, Not Optimization

The setpoint must be a *value*, not a *condition*. In particular, it is not possible to specify an extremal condition on the output (such as "the greatest possible success rate" or "the shortest possible response time").

Feedback must not be confused with an optimization scheme. It has no notion of finding the "best" settings. Instead, feedback is a *control* mechanism: it will find (and maintain) the appropriate process inputs to produce a specific desired output, even in the presence of changing external conditions.

PART II
Practice

CHAPTER 7

Theory Preview

There exists a beautiful and rather deep theory of feedback systems, which we will sketch in Part IV. Yet for most of the applications that we are interested in, this theory is not strictly required. Moreover, the theory makes several assumptions that are not necessarily fulfilled by computer systems and is therefore not even fully applicable.

Nevertheless, the theoretical description yields several useful terms and concepts that are pervasive in all of control theory. In this chapter, we will summarize the most important of those ideas so that we can use them in the sequel. At this point, we will skip most motivation and justification—if you want to know more, please refer to Part IV.

Frequency Representation

The classical theory of feedback systems is based on a mathematical operation (the *Laplace transform*) that allows us to express any function of *time t* as a function of the (complex) *frequency s*. The two representations are completely equivalent, and we can freely transform back and forth between the time domain and the frequency domain.

The Laplace transform is not universally applicable. It applies only to systems whose time evolution is described by linear, time-invariant differential equations. Many systems in the physical world are in this category; in particular this is true for many of the mechanical, electrical, and thermal assemblies for which classical feedback theory was originally developed.

The Transfer Function

In the frequency representation, the effect that a system has on its input is encapsulated in the system's *transfer function*. If the input is given by $u(s)$ in the frequency representation, then the output $y(s)$ of the system is given simply by

$$y(s) = G(s)u(s)$$

where $G(s)$ is the system's transfer function in the frequency domain. The output $y(s)$ of the process can now be transformed back to obtain the actual behavior in the time domain.

If the behavior of the system in the *time* domain is known (usually in form of a differential equation), then the Laplace transform can be used to find an explicit expression for the transfer function. Even if no differential equation is available, one can often use experimental results to obtain a phenomenological transfer function (see Chapter 8).

Block-Diagram Algebra

In the frequency representation, then, the effect of a system acting on an input is given simply by multiplying the input with the transfer function. Obviously, this process can be repeated. To let a second system $(H(s))$ act upon the output of the first $(y(s) = G(s) u(s))$ amounts to a further multiplication:

$$z(s) = H(s)G(s)u(s)$$

Transfer functions are merely functions, so one can operate with them as with numbers. But because transfer functions also contain all information about the behavior of the systems that they describe, this means that in frequency space one can *operate with systems as if they were numbers*. In particular, one can establish "algebraic" rules that can be used to combine several components together to form more complicated assemblies. Figure 7-1 shows the three most basic such rules.

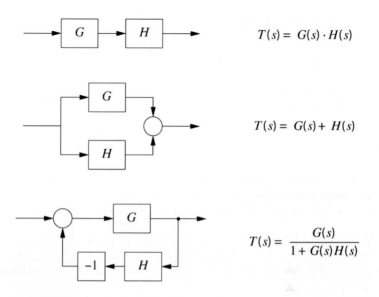

$$T(s) = G(s) \cdot H(s)$$

$$T(s) = G(s) + H(s)$$

$$T(s) = \frac{G(s)}{1 + G(s)H(s)}$$

Figure 7-1. The three most important operations of block-diagram algebra: composition in sequence (top), addition in parallel (center), and the negative feedback loop (bottom).

PID Controllers

The most common type of controller is the PID controller, which combines proportional, integral, and derivative action. In the time domain, its output (the control action) when acting upon a tracking error (its input) is given by (see Chapter 4)

$$u(t) = k_p e(t) + k_i \int_0^t e(\tau) \, d\tau + k_d \frac{de(t)}{dt}$$

This expression can be transformed to the frequency domain, yielding the following transfer function for the PID controller:

$$K(s) = k_p + \frac{k_i}{s} + k_d s$$

This transfer function is so compact that one often (in a figure, for instance) represents the integral term by the fraction k_i/s and the derivative term by $k_d s$.

Poles of the Transfer Function

The transfer function of a system encapsulates that system's entire dynamics, but we do not need to evaluate the entire transfer function to obtain information about the most dominant modes of behavior. For that, it is sufficient to know the locations of the transfer function *poles*.

Transfer functions tend to be rational functions (one polynomial divided by another):

$$H(s) = \frac{b_m s^m + b_{m-1} s^{m-1} + \cdots + b_0}{s^n + a_{n-1} s^{n-1} + \cdots + a_0}$$

The poles of $H(s)$ are those values of s for which the denominator vanishes. At those positions, $H(s)$ will become infinite, and we say that $H(s)$ has a pole at that point.

The "frequency" $s = x + iy$ is in general a *complex* number, so poles can exist within the entire complex plane. The location of the pole within the complex plane determines the corresponding behavior of the system as follows (see Figure 7-2):

- A pole in the right half-plane (that is, with positive real part: $x > 0$) will correspond to an unstable mode, which grows over time.
- A pole in the left half-plane (with $x < 0$) corresponds to a stable mode, which diminishes over time.
- Poles on the real axis (with vanishing imaginary part: $y = 0$) indicate nonoscillatory modes. Depending on the sign of the real part, the corresponding mode grows or shrinks monotonically over time.
- Poles that have imaginary parts always occur in complex conjugate pairs (that is, to each pole at $x + iy$ there exists a complex conjugate pole at $x - iy$) and describe oscillatory behavior. The larger the magnitude of the imaginary part, the higher will be the frequency of the oscillation. The amplitude of the oscillation increases or diminishes over time, depending on the sign of the real part.

Knowing *only* the positions of the poles of $H(s)$ allows us to determine whether the system will be stable or not and whether it does have a tendency to oscillate. For this reason, schemes exist to trace out the locations of poles while some parameter (such as the controller gain) is varied. (See Chapter 24.)

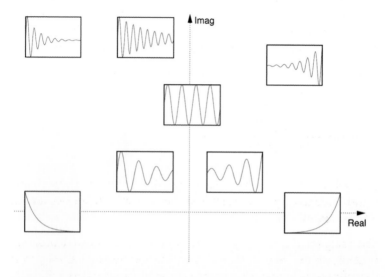

Figure 7-2. The position of a pole determines the corresponding dynamic behavior. Poles in the left half-plane lead to stable behavior; poles in the right half-plane are unstable. Poles on the real axis correspond to non-oscillatory (monotonic) behavior; poles off of the real axis are oscillatory. (The bottom half-plane is not shown because the behavior is symmetrical to the top half-plane.)

Process Models

Ideally, we have a good model of a system's behavior, typically in the form of a differential equation. If this is not the case, then we need to formulate a *phenomenological* process model, which will be based on the results of experiments undertaken with the purpose of understanding the system's dynamic behavior. Depending on our understanding of the system and also on the accuracy of the experiments, the process models developed in this way may be more or less sophisticated.

For the type of systems that we are most concerned about, there is often not much information available and the experimental results tend to be poor. Under such circumstances, it doesn't make sense to build complicated process models because simple ones will be sufficient to capture all the information available. Yet because we care about the model's representation not only in the time domain (where the experiment is carried out) but also in the frequency domain (where calculations are performed), we will try to find models that are simple in *both* domains. Luckily, simple time-domain behavior typically leads to simple transfer functions as well. (We will see examples in Chapter 8.)

Measuring the Transfer Function

If we have a good theoretical model for the system under consideration, then we can derive the transfer function directly from the model by calculating the Laplace transform of the differential equation that describes the system dynamics. More often than not, however, there won't be a good analytical model. In those cases, we will have to *measure* the transfer function in a process known as *system identification*. Even if we have a good model, we will still need to perform some measurements to "fit" the model's parameters.

There are basically two different questions we need to ask.

Static input/output relation:
>   If an input change of a certain size is applied, what's the size and direction of the ultimate change in process output?

Dynamic process response:
>   If an input change is applied suddenly, how long does it take for the system to respond?

These are the basic questions we want to answer through observations. The answer to the first one is captured in the static *process characteristic*, the answer to the second in the dynamic *process reaction curve* or *plant signature*.

All measurements are done in an *open-loop* arrangement, and *without* a controller:

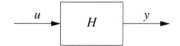

In that way, we can adjust the input in an arbitrary fashion as desired, so that we observe only the response of the system or plant alone.

Static Input/Output Relation: The Process Characteristic

The static process characteristic provides us with some basic but essential information. Obtaining it seems simple enough: apply a steady input value, wait until the system has settled down, and record the output. A typical graph might look like the one in Figure 8-1. There is a minimal control input that is required to bring about any change, and for large inputs the system begins to saturate and no longer follows the input faithfully.

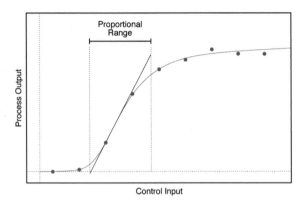

Figure 8-1. Typical process characteristic, showing both the proportional range and the areas of input saturation.

The most important feature of this curve is its local slope, which in this context is also known as the *process gain*. Specifically:

- The *magnitude* of the process gain provides information about the size or strength of the control actions that will be needed to bring about significant change in process output.

- The *sign* of the process gain provides information about the directionality of the input/output relation; if it is positive, then a regular control loop will work. But if the process gain is negative, then we must use an inverted loop arrangement (see Chapter 5).

- If the process gain undergoes drastic *change* over the typical operating range (so that control actions of different strength are needed to bring about comparable change for different input values), then the system is harder to control and we may need to consider *gain scheduling* (see Chapter 11).

Ideally, we'll be able to operate our system in the linear region, where the process output is more or less proportional to the control input (as indicated in Figure 8-1).

Practical Considerations

The simple description of the measurement process omits several practical considerations.

- How long should we wait—after making the input change—for the system to reach its new steady state, so that we can take our measurement?

- Will the system reach a steady state at all, or will the output keep growing unless stopped? (The cache hit rate is an example for a *self-regulating process* that naturally reaches a steady state. The opposite is an *accumulating process*, such as water flowing into a tank: once the input valve is open, the level in the tank will continue to increase.)

- How repeatable are the measurements? If we run the same experiment multiple times, how much difference is there in the observed outputs? (This is a measure for the amount of noise in the system.)

- If we perform the experiment with input values *increasing* from data point to data point, and then run it again with input values *decreasing*, do we find the same results? If not, this is a sign of hysteresis in the plant.

- Will we even be able to perform extensive experimentation? On production systems, it may not be possible to change the control input in a random fashion.

We can see that, in many cases, circumstances are such that we will have to make do with very small data sets indeed—possibly consisting of just a handful of points. (In Chapter 14, we will discuss a system where each data point requires a full day before a measurement can be taken.)

Nevertheless, we must be sure to obtain the two crucial bits of information about the process: the *sign* of the process gain and an approximate estimate for its *magnitude*. Without those bits of knowledge, we can't proceed.

Dynamic Response to a Step Input: The Process Reaction Curve

To measure the dynamic response, all we really have to do is switch the system on and see what happens! The system should be at rest initially (with zero input). We then apply a sudden input change and record the development of the output value over time. To obtain a good signal-to-noise ratio, the input step should be large—and we should probably repeat the whole process a few times with different input amplitudes. (Remember that this is done on the plant alone, without feedback and without a controller.)

It is possible in theory to extract all information about the transfer function from the step response, but in practice we are usually most interested in just a few essential parameters. For us, the following three are the most important (compare Figure 8-2).

Process gain K:
> This is the ratio between the value of the applied input signal and the value of the final, steady-state process output after all transients have disappeared. If the process is in the proportional range, then the process gain will be independent of the input value (thus, an input signal that is twice as large will lead to an output that it also twice as large).

Time constant T:
> The time it takes for the process to settle to a new steady state after experiencing a disturbance. The time constant is usually defined as the time it takes the process to reach about two thirds (or $1 - \exp(-1) \approx 0.63$) of its final value. (The process output approaches the steady state asymptotically, so in principle the time required to reach 100 percent of the final value is infinite.)

Dead time τ:
> Some processes exhibit a measurable delay until an input change begins to affect the output. In physical systems, such delays are usually due to transport phenomena (like liquid flowing through a pipe, or heat through a conductor, before reaching a sensor).

Some important tuning methods are based on those parameters alone (see Chapter 9).

Practical Aspects

Similar considerations apply in the case of experiments to determine the dynamic response as when attempting to obtain the static process characteristic, as discussed previously. In particular, we must be able to apply a step input in a controlled fashion and then prevent any further changes in input until the system has reached its steady state. In the case of accumulating processes that do not settle to a steady state, we must decide how long we can run them before exceeding the system's buffering capacity.

The primary reason for running experiments of this kind is to obtain enough information for controller tuning. In Chapter 9, we will discuss the Ziegler–Nichols tuning method, which attempts to make do with only minimal process knowledge.

Process Models

We can also attempt to formulate an analytical model for the transfer function and then "fit" it to the data obtained from the step-input experiment (also known as a "bump test"). In the absence of knowledge about the process internals, we will choose models that are both simple and convenient and also do a reasonable job of replicating the observed behavior. These models tend to be parameterized by the same three basic quantities already introduced: process gain K, time constant T, and delay τ. Such models are, of course, only phenomenological; their predictions need to be taken with a grain of salt, since they are not justified by any theoretical arguments. (If we have knowledge about the process internals, then we should of course take it into account when formulating a model.) Examples of some of the most frequently encountered behaviors are discussed next.

Self-Regulating Process

This is the easiest case: in response to a step-like input, the system simply approaches the steady state, possibly after a delay, but without overshoot or oscillations (see Figure 8-2). Because they eventually settle to a steady state, such processes are called *self-regulating*.

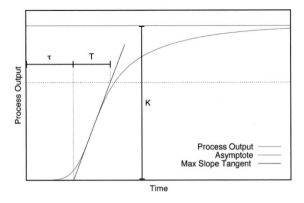

Figure 8-2. Typical process reaction curve (plant signature) for a self-regulating process.

One can obtain a rough estimate for the three parameters by using the geometric construction shown in Figure 8-2, where a tangent is drawn through the inflection point (the point with greatest slope) of the process reaction curve. The intersections of this tangent line with $y = 0$ and $y = 0.63$ are then used to determine the two time scales τ and T. The process gain K is found from the value of the process output in the long-time limit (divided by the amplitude of the input step change).

Alternatively, one can estimate the parameters by fitting an appropriate model. The following model (in the time domain) is often used to describe the step response for this type of process (see Figure 8-3):

$$f_1(t) = \begin{cases} K\left(1 - e^{-(t-\tau)/T}\right) & \text{for } t > \tau \\ 0 & \text{otherwise} \end{cases}$$

where τ is the dead time and T is the time constant. This model is chosen mainly because it has a particularly simple transfer function in the frequency domain:[1]

$$H_1(s) = \frac{K}{1+sT} e^{-s\tau}$$

1. To find the step response, multiply the transfer function $H(s)$ by the frequency representation of the unit step, which is $1/s$, and then transform the resulting expression back into the time domain. See Chapter 20 for a worked example.

One problem with this model is that the slope of $f_1(t)$ does not vanish as $t \to \tau$. So instead we can use the following, more complex model for the step response that displays a little "foot" for small t (see Figure 8-3):

$$f_2(t) = \begin{cases} K\left[1 - \left(1 + \dfrac{t-\tau}{T}\right)e^{-(t-\tau)/T}\right] & \text{for } t > \tau \\ 0 & \text{otherwise} \end{cases}$$

Although it is more complex in the time domain, in the frequency domain this model is still fairly simple:

$$H_2(s) = \frac{K}{(1+sT)^2}e^{-s\tau}$$

Figure 8-3 shows the step response for both models. At first it may appear as if the two models are really rather different, but by changing the parameters T and τ the two models can be made to look quite similar. The figure shows both $f_1(t)$ and $f_2(t)$ with the parameter values $T = 1$ and $\tau = 2$, as well as the simple model $f_1(t)$ but with parameters $T = 1.67$ and $\tau = 2.53$ (indicated as $f_1^*(t)$ in the figure). With this choice of values, the simple model $f_1(t)$ begins to look very much like the more complex model $f_2(t)$, except for a short time at the beginning.

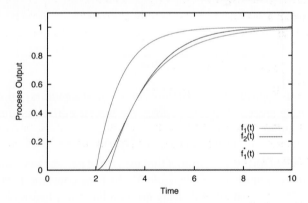

Figure 8-3. Different theoretical models for the process reaction curve of a self-regulating process. Seemingly different analytical models can lead to similar curves if the parameters are chosen appropriately.

We should take two things away from this exercise. Unless we have good reasons to choose a more complicated model, we might as well stay with a simple one, since it is (within experimental accuracy) likely to be almost as good a description of the process as a more complex one. Furthermore, we should not put too much weight on the "fitted" parameter values obtained in this way, since a seemingly small change in model can lead to rather significant changes in parameter values.

Accumulating Process

For *integrating* or *accumulating processes*, the step-input response does not settle to a steady state; instead, it continues to increase. This type of process primarily describes queueing situations and other scenarios where incoming items "pile up" until they are being handled. The models from the previous section are obviously not suitable, so a different model is needed (see Figure 8-4).

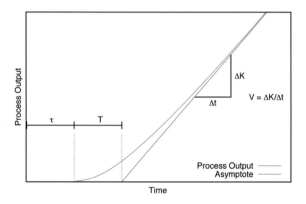

Figure 8-4. Typical process reaction curve (plant signature) for an accumulating process. Without control, the process output continues to grow with time.

We can again formulate a model that uses three parameters—namely, the *velocity gain* or *integrating gain* V, the time constant T, and the delay τ. Time constant and delay are familiar from before, but the velocity gain V must not be confused with the static gain K. The velocity gain V is a measure of the final *growth* in output, and we can obtain it from the slope of the asymptote (as shown in Figure 8-4). Accordingly, its dimension is process output over time. (In contrast, the gain K is a measure of the final process output itself and has the same dimension as the output signal.)

$$f_1(t) = \begin{cases} V\left[(t-\tau) - T\left(1 - e^{-(t-\tau)/T}\right)\right] & \text{if } t > \tau \\ 0 & \text{otherwise} \end{cases}$$

In the frequency domain, this model has the form

$$H_1(s) = \frac{V}{s(1+sT)}e^{-s\tau}$$

The most important feature of accumulating processes is, of course, the asymptotic growth in process output. Hence we can often neglect the internal dynamics, which are determined by T. This leads to a simplified model (see Figure 8-5) with the following step response in the time domain:

$$f_2(t) = \begin{cases} V(t-\tau) & \text{if } t > \tau \\ 0 & \text{otherwise} \end{cases}$$

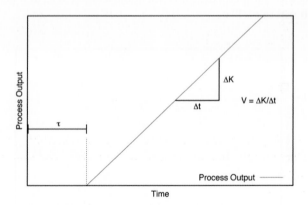

Figure 8-5. A simplified process model for accumulating processes, neglecting the plant's internal dynamics.

The model itself has the following transfer function in the frequency domain:

$$H_2(s) = \frac{V}{s} e^{-s\tau}$$

The delay τ is the time after the step input at which the process output first becomes nonzero. (Of course, the delay may be zero.) The velocity gain V must be determined from the slope of the curve, as shown in the figure.

Self-Regulating Process with Oscillation

Many mechanical or electrical devices exhibit a behavior that is more complicated than the ones just discussed. Such systems do not simply approach a new steady state value in response to an external disturbance but rather exhibit damped *oscillations*. In other words, in response to a step input, they overshoot the final value initially and then continue to oscillate around it for some time (see Figure 8-6). Think of a mass on a spring whose free end is suddenly moved to a different position: unless its motion is restricted (by being submerged in honey or otherwise damped), the mass will not just creep to its new position; it will begin to oscillate.

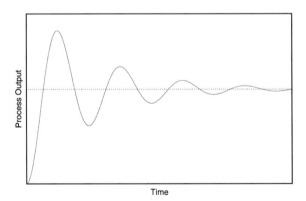

Figure 8-6. Process reaction curve for a self-regulating process with oscillations.

A common step-response model exhibiting oscillations is

$$f(t) = K \left[1 - e^{-\zeta \omega_0 t} \left(\zeta \frac{\omega_0}{\omega} \sin \omega t + \cos \omega t \right) \right]$$

with the frequency domain representation

$$F(s) = \frac{K \omega_0^2}{s^2 + 2\zeta \omega_0 + \omega_0^2}$$

Here ω_0 is the natural frequency of the system, ζ is the damping factor (controlling how quickly the amplitude of the oscillations diminishes), and $\omega = \omega_0 \sqrt{1 - \zeta^2}$ is the frequency of the damped oscillations.

Although extremely common among mechanical and electrical devices, this process model is rare in computer systems or industrial processes.[2]

Non-Minimum Phase System

Systems with this ungainly name have the perverse characteristic that their initial response to a control input is in the opposite direction from the input! (See Figure 8-7.)

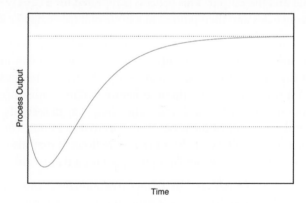

Figure 8-7. Process reaction curve for a non-minimum-phase system.

2. The details are discussed in every book on control theory. See Appendix D for some suggestions.

This is not as far-fetched as it may sound. For instance, think of a compute server, where the input is the number of active CPU instances and the output is the average query response time. If the process of activating additional CPUs takes cycles away from the already active CPUs, then the query response time will suffer while those new CPUs are being activated. More generally, this kind of behavior can occur whenever the system incorporates *two* distinct processes (with different time constants) that move the output in opposite directions. Systems exhibiting this kind of inverse response are difficult to control and require special techniques.

Other Methods of System Identification

The appeal of step input methods is their simplicity. For situations where we have a need for higher accuracy (and where we have additional understanding of the system dynamics), there are other, more accurate, methods. For example, rather than applying a step input, we can (at least in principle) apply a sinusoidal input signal. If the system under investigation is, in fact, linear, then its output to such an input will also have the shape of a sine wave with the *same* frequency but perhaps with a different amplitude and phase. We would therefore apply a sine input with a given frequency, wait until all transient behavior has died away, and then compare the amplitude and phase of the input and the output. This process is repeated for a variety of (input) frequencies and then plotted in a Bode plot (see Chapter 25).

This method has great theoretical appeal but can be difficult to apply outside the electronics lab, mostly because it is time consuming. For each of the (many) frequencies that need to be tried, one must wait until all transients have disappeared before making a measurement. This does not matter much if transients disappear in seconds, but if the typical time scale of the process is measured in minutes or hours then the process will clearly become very tedious. It may also not be feasible to generate a sinusoidal input signal for a real-world process.

Still another set of methods is based on correlation functions. An input signal is applied, and the correlation function between the input signal and the output signal is calculated. From this information, the system's transfer function can be calculated.[3]

---

3. The details are far beyond the scope of this book. Some accessible information can be found in *The Art of Control Engineering* by K. Dutton, et al. (1997).

CHAPTER 9

PID Tuning

Although the functional form of a PID controller is fixed, the gain parameters k_p, k_i, and k_d are initially undetermined. To obtain a concrete implementation, we *must* select values for these parameters. At the same time, we are *free* to choose values that will lead to the most desirable behavior, given the situation and our objectives.

Finding appropriate values for the controller gains ("tuning" the controller) can be a frustrating exercise:[1] with two (for a PI controller) or even three (for a PID controller) parameters, the number of possible combinations to try out is very large. Moreover, it is often difficult to predict intuitively what effect an increase or decrease of any one of the parameters will have on the performance of the entire feedback loop. Some sort of guidance is therefore highly desirable.

If a good analytical model of the process is available, then root locus techniques (Chapter 24) can be extremely helpful. But if no analytical expression for the transfer function is known, then we must resort to measuring the dynamic response of the system and base our tuning strategy on the experimental results. The Ziegler–Nichols rules are a classic set of heuristics that require only a little information about the process. We can go a step further and first "fit" a phenomenological transfer function model to the experimental data (Chapter 8). That model is then used to derive suitable values for the controller gains

---

1. Some studies have found that over 95 percent of industrially installed controllers are of the PID type—and that 80 percent of them function poorly, often because of improper tuning.

analytically. Methods taking this approach include the Cohen–Coon and the modern AMIGO methods.

Tuning Objectives

The first goal of controller tuning is *stability*—unless we can be sure that the system won't blow up, nothing else matters much. Once we have established stability boundaries for the parameters, we can attempt to find the best values within those boundaries in order to achieve the desired performance of the overall, closed-loop system.

Control systems can be optimized for different behaviors depending on the specific situation. Generally, this involves making typical engineering trade-offs between different desirable properties: fast systems are more susceptible to noise and oscillatory behavior; systems that are sluggish may provide better steady-state accuracy and robustness.

The most important questions for the performance of the overall, closed-loop system are as follows.

1. *Is a non-vanishing tracking error in the steady state acceptable?* For the overall system, a persistent error in the steady state is usually not acceptable (suggesting the use of an integral term in the controller). However, a complex control system may contain subsystems for which quick responses are more important than tracking accuracy. (Integral terms tend to slow the response down.)

2. *Is oscillatory behavior acceptable, and how quickly do the oscillations decay?* Oscillatory behavior is often undesirable by itself (just imagine a car's cruise-control system subjecting the passengers to such an experience). Furthermore, oscillatory systems necessarily *overshoot* the final settling value initially, which may be prohibited because it would violate some external constraint. That being said, oscillatory systems respond faster than overdamped systems.

3. *How quickly does the system have to respond to input changes?* The response time is determined by the duration of one period (for oscillatory systems) or by the time until the system reaches about two-thirds of its new steady-state value (for non-oscillatory systems). The system cannot respond faster than its dominant time scale.

4. *Must the system be robust to noise?* Noise is a high-frequency disturbance. To suppress its influence, the system needs to be relatively sluggish. This will often rule out derivative control, but it also requires longer response times overall. Systems that are robust to noise respond more slowly.

Specific performance requirements can be expressed in terms of various properties of the step response of the closed-loop system. For instance, we may require that the closed-loop system must have a "rise time" t_r of less than 2.5 seconds to ensure sufficiently speedy response. (See Figure 9-1; the "settling time" is the time until the amplitude of the oscillations has fallen to less than 5 percent of the steady state.)

In addition to the customary requirements just enumerated, further questions may arise from time to time. For instance, if very high tracking accuracy is required, then a nested control loop (see Chapter 11) may be a good idea.

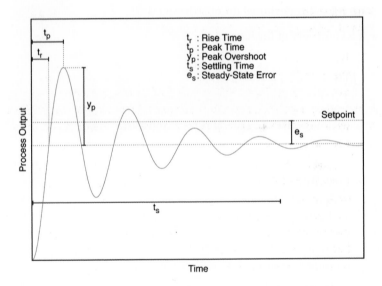

Figure 9-1. The step response of an oscillatory system together with various quantities that can be used to describe its behavior.

All the standard tuning "rules" (such as the Ziegler–Nichols and other methods discussed later in this chapter) make implicit choices about the desired accuracy and response time. These choices are intended to lead to acceptable performance for most practical applications, but they might well require augmentation on a case-by-case basis to deal with special situations.

General Effect of Changes to Controller Parameters

We can make some general statements about the typical effects that changes to the controller gains have on closed-loop performance. These observations can be useful when making manual adjustments to the values obtained from one of the systematic tuning methods described later in this chapter.

In general, increasing the controller gains leads to a speedier response but also tends to make the system less stable. This is not true for the derivative term: increasing the derivative gain leads to both greater speed and greater stability, *provided* that the signal is sufficiently free of noise. A nonzero integral gain is usually necessary to avoid a steady-state error (proportional droop, Chapter 4).

For a controller of the form

$$K(s) = k_p + \frac{k_i}{s} + k_d s$$

we can summarize the general rules as follows:

- *Increasing k_p:*
  — increases speed
  — decreases stability
  — enhances noise
- *Increasing k_i:*
  — decreases speed
  — decreases stability
  — reduces noise
  — eliminates steady-state errors more quickly
  — increases the tendency to oscillate
- *Increasing k_d:*
  — increases speed

— increases stability

— strongly enhances noise

These observations about the effects of changing the controller gains are often true, but not always. There are plants or processes that show different behavior—for example, one can find "conditionally stable" processes that become more stable when the proportional gain is increased in a closed-loop configuration.

Ziegler–Nichols Tuning

The Ziegler–Nichols tuning rules[2] for PID controllers are a set of simple heuristics intended to give adequate performance in a wide variety of situations. An essential aspect of Ziegler–Nichols tuning is that no knowledge of the plant's transfer function is required: the rules are expressed entirely in terms the plant's step-input response.

The Ziegler–Nichols rules are primarily intended for self-regulating processes, which eventually reach a steady state in response to a disturbance (Chapter 8). The method is similar to the one described in Chapter 8 to measure the dynamic response. Initially, the system is at rest, and then the response to a sudden setpoint change (in an open-loop configuration and without a controller) is observed. A tangent is fitted to the inflection point of the response curve (the point of greatest slope), and the intersection of the tangent line with the coordinate axes yields estimates for two parameters: τ and λ (see Figure 9-2).

Once τ and λ are known, numerical values for the parameters in a PID controller can be found from the formulas included in Figure 9-2. Notice that Ziegler–Nichols rules are usually quoted in a way that assumes the controller $K(s)$ is of the form

$$K(s) = k \left[1 + \frac{1}{sT_i} + sT_d \right]$$

2. This section describes the *step-response* variant of the Ziegler–Nichols method; there is also the so-called *frequency-response* variant. When using the frequency technique, integral and derivative controls are disabled and then the controller gain is increased "until the system begins to exhibit sustained, stable oscillations." The controller gain values are then expressed in terms of the frequency and gain of this oscillation. Although important for electrical devices, it is hard to see how this method can be applied to general processes.

For controllers in the form $K(s) = k_p + k_i/s + k_d s$, it is necessary to use the following conversion formula:

$$k_p = k \qquad k_i = \frac{k}{T_i} \qquad k_d = kT_d$$

The primary appeal of the Ziegler–Nichols method is its simplicity. The formulas for the controller gains are elementary, and the experimental procedure is simple and relatively fast. The reason is that it does not require the experiment to continue until the plant has reached its steady state—it is sufficient to wait until the curve exhibits an inflection point. However, the results will rarely be optimal, though they often turn out to be "good enough" in practice. Other times, they merely provide a starting point.

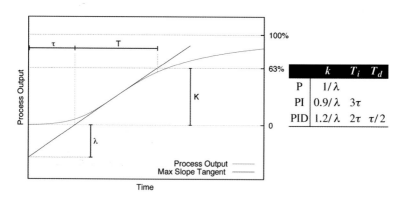

Figure 9-2. The Ziegler–Nichols tuning method.

Semi-Analytical Tuning Methods

The Ziegler–Nichols rules are pure heuristics that were developed largely through empirical observations on real systems but without theoretical justification. A more analytical approach, which also utilizes more process information, involves first fitting a model to the step response and then moving the poles of the resulting transfer function to the desired locations (see Chapter 23). The resulting expression is then solved for the controller gains in terms of the model parameters. Because the models used for this purpose are sufficiently simple (in essence, they are the models we encountered in Chapter 8), one can express the result as closed formulas that require only "plugging in" of the experimental parameters.

Table 9-1 gives the results for two such methods—the classical Cohen–Coon method and the more modern AMIGO method.[3] Both methods employ the same model introduced in Chapter 8 to describe nonoscillatory self-regulating processes:

$$F(s) = \frac{K}{1+sT} e^{-s\tau}$$

Table 9-1. Cohen–Coon and AMIGO tuning formulas.

		k	T_i	T_d
Cohen–Coon	P	$\frac{1}{K}\left(1 + \frac{0.35\theta}{1-\theta}\right)\frac{T}{\tau}$		
	PI	$\frac{0.9}{K}\left(1 + \frac{0.92\theta}{1-\theta}\right)\frac{T}{\tau}$	$\frac{3.3 - 3.0\theta}{1 + 1.2\theta}\tau$	
	PID	$\frac{1.35}{K}\left(1 + \frac{0.18\theta}{1-\theta}\right)\frac{T}{\tau}$	$\frac{2.5 - 2.0\theta}{1 - 0.39\theta}\tau$	$\frac{0.37(1-\theta)}{1 - 0.81\theta}\tau$
AMIGO self-reg	PI	$\frac{1}{K}\left[0.15 + \left(0.35 - \frac{\tau T}{(\tau + T)^2}\right)\frac{T}{\tau}\right]$	$\left[0.35 + \frac{13T^2}{T^2 + 12\tau T + 7\tau^2}\right]\tau$	
	PID	$\frac{1}{K}\left[0.2 + 0.45\frac{T}{\tau}\right]$	$\frac{0.4\tau + 0.8T}{\tau + 0.1T}\tau$	$\frac{0.5T}{0.3\tau + T}\tau$
AMIGO accumul	PI	$\frac{0.35}{V\tau}$	13.35τ	
	PID	$\frac{0.45}{V\tau}$	8τ	0.5τ

The AMIGO method also has a variant for accumulating processes, as described by the following model (also familiar from Chapter 8):

$$F(s) = \frac{V}{s} e^{-s\tau}$$

We must obtain values for K (or V), T, and τ from comparison with a step-response experiment as described earlier. The Cohen–Coon method uses the following dimensionless combinations of T and τ:

$$\theta = \frac{\tau}{\tau + T}$$

---

3. The formulas given here follow the presentation of *Advanced PID Control* by Karl J. Ångström and Tore Hägglund (2005).

This quantity is sometimes known as the "controllability ratio." For processes with a delay that is large compared to the process-internal time constant, θ approaches 1. Such processes are harder to control than processes with small θ.

Practical Aspects

The impressive appearance of the formulas in Table 9-1 should not obscure the fundamental limits of this approach. First of all, generic formulas cannot take problem-specific requirements into account. They were derived to achieve specific closed-loop performance characteristics, which may or may not provide the best balance of speed and stability for a particular application.

More importantly, these methods silently assume that the various parameters (K, V, T, and τ) can be observed with satisfactory accuracy (to within 1 percent for Cohen–Coon and AMIGO!). More often than not, this assumption is not justified in practice. Besides the obvious culprits (imperfect experimental setups, variations between experimental runs, and noise in the data), there is a more fundamental problem lurking here: the simple-lag-with-delay model underlying all these methods may not be particularly suitable for a given process. We saw in Chapter 8 that different models may fit the same data comparably well yet lead to quite different values for the parameters.

A particular concern is the "delay" parameter τ, since none of the methods give meaningful results if the observed delay vanishes. For industrial processes that have a lot of "inertia," it is reasonable to expect that all responses are gradual and exhibit an apparent delay (as in Figure 9-2), but for other types of system the importance attached to τ seems overstated.

The practical advice is that, when given a set of experimental observations, try to make the best parameter estimates "in the spirit" of the methods described. Because the three methods make slightly different assumptions and attempt to optimize performance in slightly different ways, using all three to calculate controller gains will provide a range of numerical values over which one can expect reasonable performance of the closed-loop system. (In this context it is noteworthy that the Cohen–Coon method results are similar to those of the Ziegler–Nichols method for small θ.)

A Closer Look at Controller Tuning Formulas

Tuning methods such as the Ziegler–Nichols method and its relatives were developed using pole-placement methods (Chapter 23), primarily to describe industrial processes. It is interesting to take a closer, deconstructive look at the results in order to understand what's going on—in particular with an eye to situations where the original assumptions are not valid.

If we plug the results of any one of the three methods presented here into the transfer function for the PID controller, we find that we can always write it in the following form (but do not confuse the controller's transfer function $K(s)$ with the process gain K):

$$K(s) = C\left[\alpha + \frac{\beta}{\tau s} + \gamma\tau s\right]$$

with numerical coefficients α, β, and γ. Only the numerical values of the coefficients differ from method to method. Typically they fall into the following ranges:

$$\alpha = 0.3, \ldots, 1.2 \qquad \beta = 0.25, \ldots, 1.0 \qquad \gamma = 0.4, \ldots, 0.5$$

For the Cohen–Coon and AMIGO methods, the factor C equals $\frac{T}{K\tau}$. For the Ziegler–Nichols method, the factor C equals $1/\lambda$, where λ is the intersection of the tangent with the vertical axis. But since the slope of the tangent is K/T and since the tangent passes through 0 at τ, we can write $\frac{1}{\lambda} = \frac{T}{K\tau}$ as in the other two methods. Therefore, the factor C is always $C = \frac{T}{K\tau}$, even for the Ziegler–Nichols method.

Finally, the process gain K is the ratio of the change in process output Δy that results from a change in control input Δu:

$$K = \frac{\Delta y}{\Delta u}$$

Pulling all the pieces together, we can write the controller's transfer function as

$$K(s) = \frac{T}{\tau}\left[\alpha + \frac{\beta}{\tau s} + \gamma\tau s\right]\frac{\Delta u}{\Delta y}$$

This result makes imminent sense. The rightmost factor $\Delta u / \Delta y$ captures the static behavior of the plant: how much the input u needs to change in order to bring about a desired change in output y. When acting on a tracking error e, this factor will yield the change in input required to bring about a static change in output that equals e in magnitude.

The rest of the transfer function consists of adjustments—to the size of the control action—that take the system's dynamic, time-dependent behavior into account. The leftmost factor T/τ measures to what extent the dynamics are dominated by lags or delays. Here it is helpful to adopt a broader notion of T and τ than "intersections on a graph." Namely, the "delay" τ is the time duration until a change in input first becomes *visible* in the output. The "time constant" T is the time it takes for a change to fully take effect. The factor T/τ informs us that a sluggish system requires stronger actions: if it takes twice as long for a control action to fully take effect, then we must apply twice as large a correction to achieve the same effect in the same time.

We can say this differently: when acting on an error e, the term $\frac{\Delta u}{\Delta y}e$ yields the control action u required to bring about a change in process output that would cancel e—*provided* the plant responds completely within the next time step. But since the plant response is stretched over a duration T, the control action needs to be increased by the same factor in order to bring about an error-canceling response in the next time step.

Conversely, if the delay doubles before an input change is visible in the output, then we must apply only half the correction because we will be applying it for twice as long (before the effect becomes visible).

Finally, the central term in brackets contains adjustments that are specific to each term in the PID controller. We find that the integral term is reduced by the apparent "delay" τ, which makes sense when one considers that the cumulative error has a tendency to build up during the time that a tracking error persists.

CHAPTER 10

Implementation Issues

Implementing a feedback loop based on a PID controller involves some low-level choices in addition to the overall concerns about stability and performance.

Actuator Saturation and Integrator Windup

In principle, there is no limit on the magnitude of the controller's output: when the controller gain is sufficiently large, the controller output u can become arbitrarily large. But it's a whole different question whether the downstream system (the "plant") will be able to *follow* this signal. It may either not have enough "power" to respond to an arbitrarily large input, or we may run into an even more fundamental limitation.

Think of a heated room. Given a high enough setting on the dial, the desired heat output from the central heating system can be very large—quite possibly *larger* than the amount of heat the heating system can actually produce. But even more dramatic is the opposite scenario in which we select a desired temperature that is lower than the current room temperature. In this case, the best the controller can do is to switch the heat off—there is no way for it to actively *lower* the temperature in the room (unless it is coupled to an air conditioning unit).

Such limitations always exist. In the case of a pool of compute servers, the maximum number of servers is limited: once they are all online, further demands from the controller will have no effect. At the other extreme, the number of active servers can never fall below zero. And so on.

In classical control engineering, the system that translates the controller output to an actual physical action is called the *actuator* (such as the motor that drives a valve). When an actuator is fully engaged (that is, either fully open or completely closed), it is said to be *saturated*. Hence the problem of a controlled system being unable to follow the controller output is known as *actuator saturation*.

Actuator saturation is something to be aware of. It places fundamental limitations on the performance of the entire control system. These limitations will not show up in an analysis of the transfer function: the transfer function assumes that all components are linear, and saturation effects are profoundly nonlinear. Instead, one must estimate the magnitude of the largest expected control signals separately and evaluate whether the actuator will be capable of following them. In simulations, too, one must be sure to model accurately the system's real-world limitations.

During production, the control system should rarely reach a saturated state (and it is probably a good idea to trigger an alert when it does). However, it is not safe to assume that "it won't happen"—because it will, and more often than one might suppose.

Preventing Integrator Windup

Actuator saturation can have a peculiar effect when it occurs in a control loop involving an integral controller. When the actuator saturates, an increased control signal no longer results in a correspondingly larger corrective action. Because the actuator is unable to pass the appropriate values to the plant, tracking errors will not be corrected and will therefore persist. The integrator will add them up and may reach a very large value. This will pose a problem when the plant has "caught up" with its input and the error changes sign: it will now take a long time before the integrator has "unwound" itself and can begin tracking the error again.

To prevent this kind of effect, we simply need to stop adding to the integral term when the actuator saturates. (This is known as "conditional integration" or "*integrator clamping.*") Like the actuator saturation that causes it, integrator windup should not happen during production, but it occurs often enough that mechanisms (such as clamping) *must* be in place to prevent its effects.

Setpoint Changes and Integrator Preloading

The opposite problem occurs when we first switch the system on or when we make large setpoint changes. Such sudden changes can easily overload (saturate) the actuators. In such cases, we may want to pre-load the integral term in the controller with an appropriate value so that the system can respond smoothly to the setpoint change. (This is known as *bumpless transfer* in control theory lingo.)

As an example, consider the server pool described in Chapter 5. The entire system is initially offline, and we are about to bring it up. We may know that we will need approximately 10 active server instances. (Ultimately, we may need 8 or we may need 12, but we know it's in that vicinity.) We also know the value of the coefficient k_i of the integral term in the controller. In the steady state, the tracking error e will be zero and so the contribution from the proportional term $k_p\, e$ will vanish. Therefore, the entire control output u will be due to the integral term $k_i\, I$. Since we know that u should be approximately 10, it follows that we should preload the integral term I to $10/k_i$.

Smoothing the Derivative Term

Whereas the integral term has a tendency to smooth out noise, the derivative term has a tendency to amplify it. That's inevitable: the derivative term responds to *change* in its input and noise consists of nothing but rapid change. However, we don't want to base control actions on random noise but on the overall trend in the tracking error. So if we want to make use of the derivative term in a noisy system, we must get rid of the noise—in other words, we need to smooth or filter it.

Most often, we will calculate the derivative by a finite-difference approximation. In this approximation, the value of the derivative at time step t is

$$D_t = \frac{de(t)}{dt} \approx \frac{e_t - e_{t-1}}{\delta t}$$

where e_t is the tracking error at time step t and δt is the length of the time interval between successive steps.

In a discrete-time implementation, recursive filters are a convenient way to achieve a smoothing effect. If we apply a first-order recursive filter (equivalent to "exponential smoothing") to the derivative term, the result is the following formula for the *smoothed* discrete-time approximation to the derivative $\overline{D}_t$ at time step t:

$$\overline{D}_t = \alpha \frac{e_t - e_{t-1}}{\delta t} + (1 - \alpha)\overline{D}_{t-1} \qquad \alpha = 0, \ \ldots, \ 1$$

The parameter α controls the amount of smoothing: the smaller α is, the more strongly is high-frequency noise suppressed.

The parameter α introduces a further degree of freedom into the controller (in addition to the controller gains k_p, k_i, and k_d), which needs to be configured to obtain optimal performance. This is not easy: stronger smoothing will allow a greater derivative gain k_d but will also introduce a greater lag (in effect, slowing the controller input down), thus counteracting the usual reason that made us consider derivative action in the first place. Nevertheless, a smoothed derivative term can improve performance even in noisy situations (see Chapter 16 for an example).

Finally, it makes no difference whether we smooth the error signal before calculating the derivative (in its discrete-time approximation) or instead apply the filter to the result (as was done in the preceding formula). When using a filter like the one employed here, the results are identical (as a little algebra will show).

Choosing a Sampling Interval

How often should control actions be calculated and applied? In an analog control system, whether built from pipes and valves or using electronic circuitry, control is applied *continuously*: the control system operates in real time, just as the plant does. But when using a digital controller, a choice needs to be made regarding the duration of the sampling interval (the length of time between successive control actions).

When controlling fast-moving processes in the physical world, computational speed may be a limiting factor, but most enterprise systems evolve slowly enough (on a scale of minutes or longer) that computational power is not a constraint. In many cases we find that the process itself imposes a limit on the update frequency, as when a downstream supplier or vendor accepts new prices or orders only once a day.

If we are free to determine the length of the sampling interval, then there are two guiding principles:

Faster is better...
In general, it is better to make many small control actions quickly than to make few, large ones. In particular, it is beneficial to respond to any deviation from the desired behavior *before* it has a chance to become large. Doing so not only makes it easier to keep a process under control, but it also prevents large deviations from affecting downstream operations.

... unless it's redundant.
On the other hand, there is not much benefit in manipulating a process much faster than the process can respond.

An additional problem can occur when using derivative control. If the derivative is calculated by finite differencing, then a very short interval will lead to round-off errors, whereas an interval that is too long will result in finite-differencing errors.

Ultimately, the sampling interval should be shorter (by a factor of at least 5 to 10) than the fastest process we want to control. If the controlled process or the environment to which the process must respond changes on the time scale of minutes, then we should be prepared to apply control actions every few seconds; if the process changes only once or twice a day, then applying a control action every few minutes will be sufficient.

(A separate concern is that the continuous-time theory, as sketched in Part IV, is valid without modifications only if the sampling interval is significantly shorter than the plant's time constant. If this condition is not fulfilled, then the discreteness of the time evolution must be taken into account explicitly in the theoretical treatment. See Chapter 26.)

Variants of the PID Controller

In Chapter 4 we saw that a PID controller consists of three terms, which in the time domain can be written as follows:

$$K(e) = k_p e(t) + k_i \int_0^t e(\tau)\, d\tau + k_d \frac{de(t)}{dt}$$

Here $e(t)$ is the tracking error, $e(t) = r(t) - y(t)$. A couple of variants of this basic idea exist, which can be useful in certain situations.

Incremental Form

It is sometimes useful simply to calculate the *change* in the control signal and send it to the controlled system as an incremental update. In the control theory literature, this is called the "velocity form" or "velocity algorithm" of the PID controller.

For a digital PID controller, the incremental (or velocity) form is straightforward. We find that the update at time step t in the control signal Δu_t is

$$\Delta u_t = k_p \left(e_t - e_{t-1} \right) + k_i e_t \delta t + k_d \frac{e_t - 2e_{t-1} + e_{t-2}}{\delta t}$$

When the derivative term is missing ($k_d = 0$), the equation for the incremental PID controller takes on an especially simple form.

The incremental form of the controller is the natural choice when the controlled system itself responds to *changes* in its control input. For instance, we can imagine a data center management tool that responds to commands such as "Spin up five more servers" or "Shut down seven servers" instead of maintaining a specific number of servers.

In a similar spirit, during the analysis phase it is sometimes more natural to think about the *changes* that should be made to the system in response to an error rather than about its state. But if the plant expects the actual desired state as its input (such as: "Maintain 35 server instances online"), then it will be necessary to insert an aggregator (or integrator) between the incremental controller and the controlled plant. This component will add up all the various control changes Δu to arrive at (and maintain) the currently desired control input u. (Combining the aggregator with the incremental controller leads us back to the standard PID controller that we started with.)

Finally, because the incremental controller does not maintain an integral term, no special provisions are required to avoid actuator saturation or to achieve bumpless transfers. Both of these phenomena arise from a lack of synchronization between the *actual* state of the controlled system and the *internal* state of the controller (as maintained by the integral term). Since an incremental controller does not maintain an internal state itself, there is only a single source of memory in the system (namely, in the aggregator or plant); this naturally precludes any possibilities of disagreement. (The control strategy described in Chapter 18 shows this quite clearly. In this case, the aggregator takes the saturation constraints of the controlled system into account when updating its internal state—without the controller needing to know about it.)

Error Feedback Versus Output Feedback

In general, the controller takes the tracking error as input. There is an alternative form, however, that (partially) ignores the setpoint and bases the calculation of the control signal only on the plant output. Its main purpose is to isolate the plant from sudden setpoint changes.

To motivate this surprising idea, consider a PID controller that includes a derivative term to control a system, operating in steady state, at a setpoint that is held constant. Now suppose we suddenly change the setpoint to a different value. If the controller is working on the tracking error $r(t) - y(t)$, then this change has a drastic effect on the derivative terms: although the plant output $y(t)$ may not change much from one moment to the next, the setpoint $r(t)$ was changed in a discontinuous fashion. Since the derivative of a discontinuous step is an impulse of infinite magnitude, it follows that the sudden setpoint change will lead to a huge signal being sent to the plant, its magnitude limited only by the actuator's operating range. This undesirable effect is known as the "derivative kick" or "setpoint kick."

Now consider the same controller but operating only on the (negative) plant output $-y(t)$. Since the setpoint $r(t)$ was held constant (except for the moment of the setpoint change), the effect of the derivative controller operating only on $y(t)$ is exactly the same as when it was operating on the tracking error $r(t) - y(t)$, but without the "derivative kick."

To summarize: if the setpoint is held constant except for occasional steplike changes, then the derivative of the output $\frac{d}{dt}(-y(t))$ is equal to the derivative of the tracking error $\frac{d}{dt}(r(t) - y(t))$ except for the infinite impulses that occur when the setpoint $r(t)$ undergoes a step change. So in this situation, basing the derivative action of a PID controller on only the output $y(t)$ has the same effect as taking the derivative of the entire tracking error $r(t) - y(t)$. Furthermore, taking the derivative of the output only also avoids the infinite impulse that results from the setpoint changes.

Similar logic can be applied to the proportional term. Only the integral term must work on the true tracking error. The most general form of this idea is to assign an arbitrary weight to the setpoint in both the proportional and derivative terms. With this modification, the full form of the PID controller becomes

$$K(e) = k_p[a\,r(t) - y(t)] + k_i \int_0^t [r(\tau) - y(\tau)]d\tau + k_d \frac{d}{dt}[b\,r(t) - y(t)]$$

with $0 \le a, b \le 1$. The parameters a and b can be chosen to achieve the desired response to setpoint changes. The entire process is known as "setpoint weighting."

The General Linear Digital Controller

If we use the finite-difference approximations to the integral and the derivative, then the output u_t at time t of a PID controller in discrete time can be written as

$$u_t = k_p e_t + k_i\, \delta t \sum_{\theta=0}^t e_\theta + \frac{k_d}{\delta t}(e_t - e_{t-1})$$

where $e_t = r_t - y_t$. If we plug the definition of e_t into the expression for u_t and then rearrange terms, we find that u_t is a linear combination of r_θ and u_θ for all possible times $\theta = 0, ..., t$.

In the case of setpoint weighting, we allowed ourselves greater freedom by introducing additional coefficients ($e_t = a\,r_t - y_t$ in the proportional term and $e_t = b\,r_t - y_t$ in the derivative term). Nothing in u_t changes structurally, but some of the coefficients are different. (Setting $a = b = 1$ brings us back to the standard PID controller.)

Generalizing even further, we have the following formula for the most general, linear controller in discrete time:

$$u_t = a_t\,r_t + a_{t-1}\,r_{t-1} + \cdots + a_0\,r_0 + b_t\,y_t + b_{t-1}y_{t-1} + \cdots + b_0 y_0$$

All controllers that we have discussed so far—including the standard PID version, its setpoint-weighted form, the incremental controller, and the derivative-filtered one—are merely special cases obtained by particular choices of the coefficients a_0, ..., a_t and b_0, ..., b_t. Further variants can be obtained by choosing different coefficients.

Nonlinear Controllers

All controllers that we have considered so far were *linear*, which means that their output was a linear transformation of their input. Linear controllers follow a simple theory and reliably exhibit predictable behavior (doubling the input will double the output). Nevertheless, there are situations where a nonlinear controller design is advisable or even necessary.

Error-Square and Gap Controllers

Occasionally the need arises for a controller that is more "forgiving" of small errors than the standard PID controller but at the same time more "aggressive" if the error becomes large. For example, we may want to control the fill level in an intermediate buffer. In such a case, we neither need nor want to maintain the fill level accurately at some specific value. Instead, it is quite acceptable if the level fluctuates to some degree—after all, the purpose of having a buffer in the first place is for it to neutralize small fluctuations in flow. However, if the fluctuations become too large—threatening either to overflow the buffer or to let it run empty—then we require drastic action to prevent either of these outcomes from occurring.

A possible modification of the standard PID controller is to multiply the output of the controller (or possibly just its proportional term) by the absolute value of the tracking error; this will have the effect of enhancing control actions for large errors and suppressing them for small errors. In other words, we modify the controller response to be

$$K_2(e) = |e(t)| \left(k_p e(t) + k_i \int_0^t e(\tau)\, d\tau + k_d \frac{de(t)}{dt} \right)$$

Because it now contains terms such as $|e|e$, this form of the controller is known as an *error-square* controller (in contrast to the *linear* error dependence of the standard PID controller). This is a rather ad hoc modification; its primary benefit is how easily it can be added to an existing PID controller.

Another approach to the same problem is to introduce a dead zone or "gap" into the controller output. Only if (the absolute value of) the error exceeds this gap is the control signal different from zero.

Both the error-square form of the controller and introduction of a dead zone are rather ad hoc modifications. The resulting controllers are nonlinear, so the linear theory based on transfer functions applies "by analogy" only.

Simulating Floating-Point Output

By construction, the PID controller produces a floating-point number as output. It is therefore suitable for controlling plants whose input can also take on any floating-point value. However, we will often be dealing with plants or processes that permit only a set of discrete input values. In a server farm, for instance, we must specify the number of server instances in whole integers.

The naive approach, of course, is simply to round (or just truncate) the floating-point output to the nearest integer. This method will work in general, but it might not yield very good performance owing to the error introduced by rounding (or truncating). A particular problem are situations that require a fractional control input in the steady state. For instance, we may find that precisely 6.4 server instances are required to handle the load: 6 are not enough, but if we deploy 7 then they won't be fully utilized. Using the simple rounding or truncating strategy in this case will lead to permanent and frequent switching between 6 and 7 instances; if random disturbances are present in the loop, then the switching will be driven primarily by random noise. (We will encounter this problem in the case studies in Chapter 15 and Chapter 16.)

Given the integer constraints of the system, it won't be possible to generate true fractional control signals. However, one can design a controller that gives the correct output signals "on average" by letting the controller switch between the two adjacent values in a controlled manner. Thus, to achieve an output of 6.4 "on average," the controller output must equal 6 for 60 percent of the time and 7 for 40 percent of the time. For this strategy to make sense, the controller output must remain relatively stable over somewhat extended periods of time. If that condition is fulfilled, then such a "split-time" controller can help to reduce the amount of random control actions.

Categorical Output

Permissible control inputs can be even more constrained than being limited to whole integers. For instance, it is conceivable that the allowed values for the process input are restricted to nonnumeric "levels" designated only by such categorical labels as "low," "medium," "high," and "very high."

These labels convey an ordering but no quantitative information. Hence the controller does not have enough information about the process (and its levels) to *select* a particular level. The only decision the controller can make is whether the current level should be incremented (or decremented) in response to the current sign and magnitude of the error.

Because it has only a few, discrete control inputs, such a system will usually not be able to track a setpoint very accurately; however, it may still be possible to restrict the process output to some predefined *range* of values. We will study an example of such a system in Chapter 18.

CHAPTER 11

Common Feedback Architectures

Sometimes a simple feedback loop is not enough. There are situations that require a combination of control elements, or even extensions beyond closed-loop feedback control. In this chapter, we will discuss some commonly occurring problems and their standard solutions.

For reference, the most general form of the basic, "textbook" control loop is shown in Figure 11-1. In addition to the familiar elements (the controller K, the plant H, and the optional return filter G), this graph also shows explicitly how *disturbances* can be included in a block diagram.

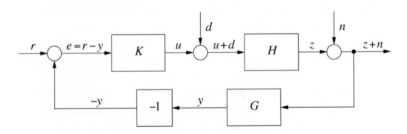

Figure 11-1. The standard feedback loop, including the effect of load disturbances d and measurement noise n.

Also shown are disturbances entering the control loop. Generally, all effects that tend to drive the system away from its desired operating point are considered disturbances. Disturbances arising within the controlled plant or system are called *load disturbances* (because they arise within the "load" that is driven by the controller); disturbances that result from imperfections in the sensors used to observe the plant output are called *measurement noise*.

In the figure, measurement noise is shown as an external signal n being added to the plant output. The load disturbance d is modeled as affecting the plant input u, but in a way that is not observable by us. This is a modeling idealization; the disturbance might actually arise *inside* the plant proper, but capturing this behavior would require us to model a plant H that itself undergoes changes in time. As long as the disturbance is not observable, we might as well treat it as a component of the plant input and treat the plant itself as static.

In physical control systems, load disturbances tend to consist of slow drifts or sudden, steplike changes in the process characteristic or the dynamic response. Measurement noise arises from imperfections of the physical sensor device and is usually a high-frequency phenomenon. For computer systems (which, incidentally, often do not even have separate "sensors"), it is more useful to distinguish disturbances primarily by their frequency spectrum: high-frequency, uncorrelated random components in a signal are called "noise" irrespective of their origin, and all other drifts and changes are viewed as load disturbances.

Changing Operating Conditions: Gain Scheduling

Sometimes a control loop needs to be operated under a wide range of different conditions. Different conditions may call for different behaviors of the controlled system and thus for different values of the controller's gain parameters. We can accommodate this possibility by setting up several different sets of values for the gain parameters and, at runtime, select the most appropriate set for the current conditions. This is known as *gain scheduling*. Ever driven a car that allowed you to choose between a "sport" and a "comfort" mode? That's gain scheduling.

The signal that indicates which set of values to use can be anything. The process input u or output y are common, but it can also be a completely external signal, such as the time of day. Some possible applications include the following:

- Using the process input u to select different modes for high and low system load.

- Using the time of day to get the system ready for "rush hour."
- Using the magnitude of the error to switch the controller to more aggressive behavior when the error has become large—for instance when controlling a queue.

The cases just described are straightforward: instead of a single set of gain parameters, we have a few, distinct such sets. At any given time, exactly one of them is active and supplies the gain parameters to the controller. We can generalize this idea and let the gain parameters float freely (see Figure 11-2) based on the value of an appropriate scheduling signal. In principle, there are no limitations on the algorithm that calculates the momentary gain values, but the dynamics of the resulting system will probably be very complex! (Moreover, a control system with arbitrarily varying gain parameters is neither time invariant nor linear, which means that standard control-theoretic methods are insufficient to analyze it.)

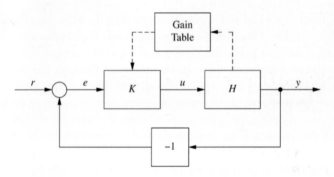

Figure 11-2. The loop architecture for a system with gain scheduling.

Gain Scheduling for Mildly Nonlinear Systems

Gain scheduling can be used for processes that have only mildly nonlinear process characteristics (see Chapter 8). Recall that when a process is linear, the shape of the dynamic response will be the same regardless of the magnitude of the input—the only difference will be the magnitude of the final steady-state value. For processes that contain nonlinearities, however, the dynamic response may be qualitatively different for different values of the input variable. If this is the case then it may be desirable or even necessary to have two (or more) sets of values for the controller gains, depending on the current operating point, to ensure satisfactory performance over the entire range of possible inputs. Basically, one of the tuning processes from Chapter 9 is executed several times with different sizes of the step input. Each experimental run leads to a different set of controller gains. In closed-loop operations, values from the appropriate set are used.

Large Disturbances: Feedforward

When feedback-controlled systems are subject to large, sudden disturbances, we can sometimes improve their performance by augmenting the "automatic" control of the feedback loop with a direct *feedforward* control strategy. Feedback works by responding to an error (the difference between the setpoint and the actual process output). This works great for small but frequent deviations. But if disturbances are large and sudden, then waiting for the error to propagate through the loop may be unsatisfactory and so a more direct corrective action is needed.

When employing feedforward, we instead "plan ahead" what the process input should be and apply this input directly, bypassing the feedback controller. This makes sense in particular if disturbances can be detected or even predicted. Scheduled setpoint changes are a good example of a situation that can be improved through feedforward. Instead of applying the setpoint change and then waiting for the system to eventually settle to a new equilibrium, we can begin to apply an appropriate change to the process before the setpoint change takes place, thus reducing the magnitude of the error that will occur. But feedforward is applicable to more than just changes in setpoint: one can also think of predictable load changes, such as rush-hour traffic patterns. In such situations, feedforward can be helpful to prepare the system for the increased load. (See Figure 11-3, top.)

That being said, feedforward should be used sparingly. The idea of calculating the control input ahead of time—and then applying it without regard to the actual output—runs counter to the very concept of feedback control. Feedforward can be useful to avoid the undesirable effects of sudden changes, but relying on it too much entails discarding the benefit of employing feedback control in the first place.

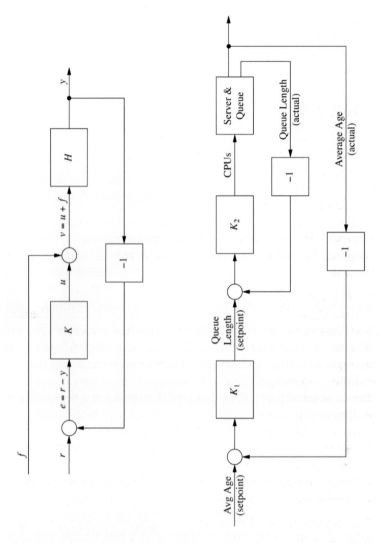

Figure 11-3. Top: The loop architecture for a system involving a feed-forward path f. Bottom: A nested or "cascaded" loop architecture. The outer loop provides the setpoint for the fast inner loop.

Fast and Slow Dynamics: Nested or "Cascade" Control

At times, we may need to control a process that has naturally "slow" dynamics and all of the associated challenges. If we are able in such situations to identify a subprocess with significantly faster responses, then we can improve the performance of the overall system by using a nested or "cascaded" control architecture; this is shown in Figure 11-3 (bottom).

We will discuss the problem in terms of the "task server" example introduced in Chapter 5. The process variable that we ultimately want to control is the average age of requests in the queue. Yet this quantity will only respond slowly because it is formed as an average over all items currently waiting. At the same time, it is necessary to control the queue length in the short term to prevent it from overflowing the buffer (this is even more of an issue in systems controlling a queue of *physical* items).

The solution is to have two nested control loops, as shown in Figure 11-3 (bottom). The inner, fast loop monitors the queue length and adjusts the number of active worker units to maintain the desired queue length. The outer, slow loop tracks the average age of waiting items and feeds the desired queue length as setpoint to the inner loop.

Nested control loops are suitable whenever two processes with very different timescales are involved. An inner loop is used to control the fast dynamics, thereby allowing for more accurate tracking of the outer, slowly responding signals. (A cascaded control system much like the one sketched here will be discussed in more detail in the case study of Chapter 16.)

Systems Involving Delays: The Smith Predictor

In systems subject to delays, *no* response to a control action is visible in the process output until the duration τ of the delay has passed. This makes such systems harder to control than lag-dominated systems, which at least exhibit an immediate (albeit partial) response. The *Smith predictor* (after O. J. M. Smith) is a control strategy intended for situations involving a pure delay.

We can think of the total dynamic response of a system subject to a delay as consisting of two parts: the delay (or dead time) of duration τ itself, during which nothing happens; and the actual dynamic response of the plant, which becomes visible in the process output once the dead time has passed. The Smith predictor assumes that we know the magnitude of the delay τ and that we also have a reasonably good model for the second part—namely, a device (or subroutine) that can reproduce the behavior of the actual plant but without the delay.

These elements are combined in a control loop as shown in Figure 11-4 (top). Concentrate first on the part of the loop shown with solid lines. These parts form a closed feedback loop consisting of the controller K and the model H_0. The feedback loop therefore controls the *model* instead of the plant, with the effect that the model output v tracks the setpoint r.

But the control signal u is also sent to the actual plant H. Because the model reproduces (at least approximately) the plant's response to control inputs, the result is that the output y of the actual plant *also* tracks the setpoint—but subject to the delay τ introduced by the plant.

In practice, no model will ever reproduce the plant's response perfectly. Furthermore, the real plant is subject to random disturbances whereas the model is not. To correct for these differences between the model and the plant, a *second* feedback loop is incorporated into the overall control architecture (shown dashed in Figure 11-4, top). The model output v is delayed by the appropriate amount (to allow the plant to "catch up"). Then the delayed model output w is compared to the output y of the plant itself. The difference $e_M = y - w$, which could be called the "modeling error," is then simply added to the return path and fed back to the controller. This modeling error e_M is subject to the full delay, yet because e_M is expected to be relatively small, the effect of the delay on the performance of the control system is less severe.

There is an alternative way to draw the loop architecture of the Smith predictor (see Figure 11-4, bottom). Although it may not be obvious, this system is equivalent to the one shown in Figure 11-4 (top).[1] Drawn in this way, we can see that the Smith predictor consists of two nested control loops. The inner loop is closed around the controller (instead of being closed around the plant as in the case of cascade control, discussed earlier). In particular, we can consider the inner loop, which comprises the actual controller K and all elements depending on the model H_0, as a regular loop element with a single input and a single output (as indicated by the dashed box in the figure). With this iden-

---

1. The easiest way to see this is to compare the controller inputs while taking into account that in both figures the plant output is denoted by y, the model output by v, and the delayed model output by w. In Figure 11-4 (top), the controller input is clearly $r - (v + (y - w))$; in Figure 11-4 (bottom), it is $r - y + (w - v)$, which is the same.

tification, the Smith predictor becomes a standard control loop but with a special controller that contains a closed loop internally.

The Smith predictor offers an approach for dealing with systems subject to delays but at the cost of increased complexity. The quality of the results depends, of course, on the quality of the model—in particular, on the accuracy with which we can determine the length of the delay τ. Finally, the Smith predictor can only be used with a plant H that is stable by itself.

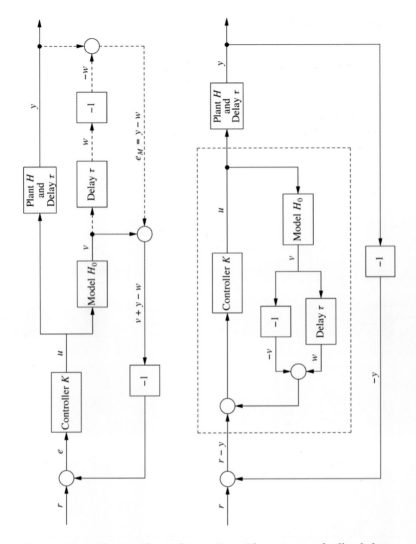

Figure 11-4. The Smith predictor. Top: The primary feedback loop, based on the model H_0 of the process, is shown with solid lines; the secondary loop, which compares the model to the actual process output, is shown dashed. Bottom: An alternative representation of the Smith predictor. Drawn this way, the Smith predictor is seen as an ordinary feedback loop but with a special type of controller (as indicated by the dashed box).

PART III
Case Studies

CHAPTER 12

Exploring Control Systems Through Simulation

The next several chapters describe—in the form of case studies—a number of control problems and also show how they can be solved using feedback mechanisms. The case studies are treated via *simulations*; the simulation code is available for download (*http://exam ples.oreilly.com/9781449361693-files/*) from the book's website.

The ability to run simulations of control systems is extremely important for several reasons.

- The behavior of control systems, specifically of feedback loops, can be unfamiliar and unintuitive. Simulations are a great way to develop intuition that is required to solve the kind of practical problems that arise in real-world control systems.

- Extensive experimentation on real production systems is often infeasible for reasons of availability and cost. Even when possible, experiments on real systems tend to be time-consuming (the time scale of many processes is measured in hours or even days); if they involve physical equipment, they can be outright dangerous.

- Implementing controllers, filters, and other components in a familiar programming language can bring abstract concepts such as "transfer functions" to life. In this way, simulations can help to make some of the theoretical concepts more concrete.

- The parts of the simulation *outside* the controlled system—that is, the components of the actual control loop proper—will carry over rather directly from the simulation to a "real" implementation.

Finally, it is unlikely that any control system will be deployed into production unless it has proven itself in a simulated environment. Simulation, therefore, is not just a surrogate activity but an indispensable step in the design and commissioning of a control system.

The Case Studies

In the following chapters, we will discuss a variety of case studies in detail. I think each case study models an interesting and relevant application. In addition, each case study demonstrates some specific problem or technique.

Cache hit rate:
How large does a cache have to be to maintain a specific hit rate? This case study is a straightforward application of many "textbook" methods. It demonstrates the use of "event time" (as opposed to "wall-clock time"—see the following section in this chapter).

Ad delivery:
How should ads be priced to achieve a desired delivery plan? We will find that this system exhibits discontinuous dynamics, which makes it susceptible to rapid control oscillations.

Server scaling:
How many server instances do we need to achieve a desired response rate? The problem is that the desired response rate is 100 percent. It is therefore not possible for the actual output to straddle the setpoint, so we need to develop a modified controller to handle this requirement.

Waiting queue control:
How many server instances do we need to manage a queue of pending requests? This example appears similar to the preceding one, but the introduction of a queue changes the nature of the problem significantly. We will discuss ways to control a quantity (such as the queue length) that is not fixed to one specific value but instead must be allowed to float. This example also demonstrates the use of nested (or "cascaded") control loops and the benefits of derivative control.

Cooling fan speed:
How fast must CPU cooling fans spin in order to achieve a desired CPU temperature? This case study shows how to simulate a physical process occurring in real time.

Graphics engine resolution:
What graphics resolution should be used to prevent memory consumption from exceeding some threshold? In this example, the control input is a nonnumeric quantity that can assume only five discrete "levels." Since PID control is not suitable for such systems, we show how to approach them using an incremental on/off controller.

Modeling Time

All processes occur over time, and control theory, in particular, is an attempt to harness the controlled system's time evolution. Simulations, too, are all about advancing the state of the simulated system from one time step to the next.

Control Time

For phenomena occurring in the physical world (mechanical, electrical, chemical processes, and so on), "time" is an absolute concept: it just passes, and things happen according to laws of nature. For computer systems, things are not as clear because computer systems do not necessarily "evolve" by themselves according to separate and fixed laws. A software process waiting for an event does not "evolve" at all while it waits for an event to occur on the expected port! (Anybody who has ever had to deal with a "hung" computer will recognize this phenomenon.) So when designing control systems for computer systems, we have a degree of freedom that most control engineers do not—namely, a choice of time.

This is a consequence of using a digital controller. Classical, analog control systems were built using physical elements (springs and dampers, pipes and valves, resistors and capacitors). These control systems were subject to the same laws of nature as the systems they controlled, and their action progressed continuously in time. Digital controllers do not act continuously; they advance only in discrete time steps (almost always using time steps of fixed length, although that is not strictly required). If the controlled system itself is also digital and therefore governed by the rules of its application software, then this introduces an additional level of freedom.

There are basically two ways that time evolution for computer systems can be designed: real time or control time.

Real or "wall-clock" time:
> The system evolves according to its own dynamics, independently (asynchronously) from control actions. Control actions will usually occur periodically with a fixed time interval between actions.

Control or "event" time:
> The system does not evolve between control events, so all time development is synchronous with control actions. In this case, we may synchronize control actions with "events" occurring in the physical world. For instance, we take action if and only if a message arrives on a port; otherwise, the system does not evolve.

When we are trying to control a process in the physical world, only the first approach is feasible: the system evolves according to its own laws, and we must make sure that our control actions keep up with it. But when developing a control strategy for an event-driven system, the second approach may be more natural. (In Chapter 13 we will see an example of "event" time; in Chapter 17 we will see how to connect simulation parameters with "wall-clock" time.)

Simulation Time

In a simulation, time naturally progresses in discrete steps. Therefore, the (integer) number of simulation steps is the natural way to tell time. The problem is how to make contact with the physical world that the simulation is supposed to describe.

In the following, we will assume that each simulation step has exactly the same duration when measured in wall-clock time. (This precludes event-driven situations, where the interval between successive events is a random quantity.) Each step in the simulation corresponds to a specific duration in the real world. Hence we need a conversion factor to translate simulation steps into real-world durations.

In the simulation framework, this conversion factor is implemented as a variable DT, which is global to the feedback package. This variable gives the duration (in wall-clock time) of a single simulation step and must be set before a simulation can be run. (For instance, if you want to measure time in seconds and if each simulation step is supposed to describe 1/100 of a second, then you would set DT = 0.01. If you measure time in days and take one simulation step per day, then DT = 1.)

The factor DT enters calculations in the simulation framework in two ways. First, the convenience functions for standard loops (see later in this chapter) write out simulation results at each step for later analysis, including both the (integer) number of simulation steps and the wall-clock time since the beginning of the simulation. (The latter is simply DT multiplied by the number of simulation steps.) The second way that DT enters the simulation results is in the calculation of integrals and derivatives, both of which are approximated by finite differences:

$$\frac{df}{dt} \approx \frac{f(t+DT)-f(t)}{DT}$$

$$\int f(t)\,dt \approx DT*sum(fs)$$

where fs = [f0, f1, f2, ...] is a sequence holding the values of $f(t)$ for all simulation steps so far.

It is a separate question how long or short (in wall-clock time) you should choose the duration of each simulation step to be. In general, you want each simulation step to be shorter by at least a factor of 5 to 10 than the dominant time scale of the system that you are modeling (see Chapter 10). This is especially important when simulating systems described by differential equations, since it ensures that the finite-difference approximation to the derivatives is reasonably accurate.

The Simulation Framework

The guiding principles for the design of the present simulation framework were to make it *simple* and *transparent* in order to demonstrate the algorithms as clearly as possible and to encourage experimentation. Little emphasis was placed on elegant implementations or run-time efficiency. The code presented is intended as a teaching tool, not as building material for production software!

In contrast to some other existing simulation frameworks for control systems, components are not specified by their frequency-domain transfer functions. Instead, algorithms are developed and implemented "from scratch" based on behavior in the time domain. This allows us to consider *any* behavior, whether or not a frequency-domain model is readily available. The purpose is to make it easy to develop one's own process models, without necessarily having to be comfortable with frequency-domain methods and Laplace transforms.

Components

All components that can occur in a simulation are subclasses of Component in package feedback. This base class provides two functions, which subclasses should override. The work() function takes a single scalar argument and returns a single scalar return value. This function encapsulates the dynamic function of the component: it is called once for each time step. Furthermore, a monitoring() function, which takes no argument, can be overridden to return an arbitrary string that represents the component's internal state. This function is used by some convenience functions (which we'll discuss later in this chapter) and allows a uniform approach to logging. Having uniform facilities for these purposes will prove convenient at times.

The complete implementation of the Component base class looks like this:

```
class Component:
    def work( self, u ):
        return u

    def monitoring( self ):
        # Overload, to print addtl monitoring info to stdout
        return ""
```

Plants and Systems

All implementation of "plants" or systems that we want to control in simulations should extend the Component base class and override the work() function (and the monitoring() function, if needed):

```
class Plant( Component ):
    def work( self, u ):
        # ... implentation goes here
```

We will see many examples in the chapters that follow.

Controllers

The feedback package provides an implementation of the standard PID controller as a subclass of the Component abstraction. Its constructor takes values for the three controller gains (k_p, k_i, and k_d; the last one defaults to zero). The work() function increments the integral term, calculates the derivative term as the difference between the previous and the current value of the input, and finally returns the sum of the contributions from all three terms:

```
class PidController( Component ):
    def __init__( self, kp, ki, kd=0 ):
        self.kp, self.ki, self.kd = kp, ki, kd
        self.i = 0
        self.d = 0
        self.prev = 0

    def work( self, e ):
        self.i += DT*e
        self.d = ( e - self.prev )/DT
        self.prev = e

        return self.kp*e + self.ki*self.i + self.kd*self.d
```

Observe that the factor DT enters the calculations twice: in the integral term and when approximating the derivative by the finite difference between the previous and the current error value.

A more advanced implementation is provided by the class AdvController. Compared to the PidController, it has two additional features: a "clamp" to prevent integrator windup and a filter for the derivative term. By default, the derivative term is not smoothed; however, by supplying a positive value less than 1 for the smoothing parameter, the contribution from the derivative term is smoothed using a simple recursive filter (single exponential smoothing).

The clamping mechanism is intended to prevent integrator windup. If the controller output exceeds limits specified in the constructor, then the integral term is not updated in the next time step. The limits should correspond to the limits of the "actuator" following the controller. (For instance, if the controller were controlling a heating element, then the lower boundary would be zero, since it is in general impossible for a heating element to produce *negative* heat flow.) This clamping mechanism has been chosen for the simplicity of its implementation—many other schemes can be conceived.

```
class AdvController( Component ):
    def __init__( self, kp, ki, kd=0,
                  clamp=(-1e10,1e10), smooth=1 ):
        self.kp, self.ki, self.kd = kp, ki, kd
        self.i = 0
        self.d = 0
        self.prev = 0
```

```
        self.unclamped = True
        self.clamp_lo, self.clamp_hi = clamp

        self.alpha = smooth

    def work( self, e ):
        if self.unclamped:
            self.i += DT*e

        self.d = ( self.alpha*(e - self.prev)/DT +
                   (1.0-self.alpha)*self.d )

        u = self.kp*e + self.ki*self.i + self.kd*self.d

        self.unclamped = ( self.clamp_lo < u < self.clamp_hi )
        self.prev = e

        return u
```

Actuators and Filters

Various other control elements can be conceived and implemented as subclasses of Component. The Identity element simply reproduces its input to its output—it is useful mostly as a default argument to several of the convenience functions:

```
class Identity( Component ):
    def work( self, x ): return x
```

More interesting is the Integrator. It maintains a cumulative sum of all its inputs and returns its current value:

```
class Integrator( Component ):
    def __init__( self ):
        self.data = 0

    def work( self, u ):
        self.data += u
        return DT*self.data
```

Because the Integrator class is supposed to calculate the *integral* of its inputs, we need to multiply the cumulative term by the wall-clock time duration DT of each simulation step.

Finally, we have two smoothing filters. The FixedFilter calculates an unweighted average over its last *n* inputs:

```
class FixedFilter( Component ):
    def __init__( self, n ):
        self.n = n
```

```
self.data = []

def work( self, x ):
    self.data.append(x)

    if len(self.data) > self.n:
        self.data.pop(0)

    return float(sum(self.data))/len(self.data)
```

The RecursiveFilter is an implementation of the simple exponential smoothing algorithm

$$s_t = \alpha x_t + (1 - \alpha)s_{t-1}$$

that mixes the current raw value x_t and the previous smoothed value s_{t-1} to obtain the current smoothed value s_t:

```
class RecursiveFilter( Component ):
    def __init__( self, alpha ):
        self.alpha = alpha
        self.y = 0

    def work( self, x ):
        self.y = self.alpha*x + (1-self.alpha)*self.y
        return self.y
```

Because all these elements adhere to the interface protocol defined by the Component base class, they can be strung together in order to create simulations of multi-element loops.

Convenience Functions for Standard Loops

In addition to various standard components, the framework also includes convenience functions to describe several standard control loop arrangements in the feedback package. The functions take instances of the required components as arguments and then perform a specified number of simulation steps of the entire control system (or control loop), writing various quantities to standard output for later analysis. The purpose of these functions is to reduce the amount of repetitive "boilerplate" code in actual simulation setups.

The closed_loop() function is the most complete of these convenience functions. It models a control loop such as the one shown in Figure 12-1. It takes three mandatory arguments: a function to provide the setpoint, a controller, and a plant instance. Controller and plant must be subclasses of Component. The setpoint argument must be a reference to a function that takes a single argument and returns a numeric value, which will be used as a setpoint for the loop. The loop provides the current simulation time step as an integer argument to the setpoint function—this makes it possible to let the setpoint change over time. These three arguments are required.

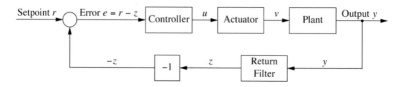

Figure 12-1. The control loop modeled by the closed_loop() function.

The remaining arguments have default values. There is the maximum number of simulation time steps tm and a flag inverted to indicate whether the tracking error should be inverted ($e \rightarrow -e$) before being passed to the controller. This is necessary in order to deal with processes for which the process output *decreases* as the process input *increases*. (This mechanism allows us to maintain the convention that controller gains are always positive, as was pointed out in Chapter 4). Finally, we can insert an arbitrary actuator between the controller and the plant, or we can introduce a filter into the return path (for instance, to smooth a noisy signal). The complete implementation of the closed control loop can then be expressed in just a few lines of code:

```python
def closed_loop( setpoint, controller, plant, tm=5000,
                 inverted=False, actuator=Identity(),
                 returnfilter=Identity() ):
    z = 0
    for t in range( tm ):
        r = setpoint(t)
        e = r - z
        if inverted == True: e = -e
        u = controller.work(e)
        v = actuator.work(u)
        y = plant.work(v)
        z = returnfilter.work(y)

        print t, t*DT, r, e, u, v, y, z, plant.monitoring()
```

In any case, we must provide an implementation of the plant and of the function to be used for the setpoint. Once this has been done, a complete simulation run can be expressed completely through the following code:

```
class Plant( Component ):
    ...

def setpoint( t ):
    return 100

p = Plant()
c = PidController( 0.5, 0.05 )

closed_loop( setpoint, c, p )
```

All convenience functions use the same output format. At each time step, values for all signals in the loop are written to standard output as white-space separated text. Each line has the following format:

1. The simulation time step
2. The wall-clock time
3. The setpoint value
4. The tracking error
5. The controller output
6. The actuator output (the plant input)
7. The plant output
8. The output of the return filter
9. Plant-specific information as returned by monitoring()

The remaining convenience functions describe a system for conducting step-response tests and an open-loop arrangement. There is also a function that conducts a complete test to determine the static, steady-state process characteristic.

Generating Graphical Output

The simulation framework itself does not include functionality to produce graphs—this task is left for specialized tools or libraries. If you want to include graphing functionality directly into your simulations, then matplotlib (*http:/matplotlib.org*) is one possible option.

The alternative is to dump the simulation results into a file and to use a separate graphing tool to plot them. The graphs for this book were created using gnuplot, although many other comparable tools exist.

You should pick the one you are most comfortable with. (As a starting point, a brief tutorial on gnuplot is included in Appendix B.)

CHAPTER 13

Case Study: Cache Hit Rate

Maintaining the "hit rate" for a cache by adjusting its size is a perfect application of feedback principles. Caches are ubiquitous and important. Their basic function is familiar and so does not distract from the control architecture, which is our primary concern. This example will allow us to discuss some design decisions that arise in the application of control principles to computer systems.

This example also serves as an interesting "metaphor" for the application of feedback principles to any form of *statistical process control*. This identification may not be immediately obvious; we will revisit it after identifying the relevant components.

Defining Components

The controlled system is a *cache*, such as a web or database cache. We will assume that the cache can hold a fixed number n of items—this will be the control or input variable. For the sake of definiteness, we will consider a cache using a "most recently accessed" protocol: if the requested document or object is found in the cache then it is returned to the requestor; if it is not found then the object is fetched from the backing store and added to the cache. If the number of items held in the cache exceeds the maximally allowed number n, then the oldest item is removed from the cache. (The specifics of the caching policy are not important to the design of the feedback loop.)

During operation, we would like to maintain a specific hit rate or success rate of requests—for instance, we may require that 70 percent of requests should be fulfilled from items held in the cache. At the same time, we want to keep the cache as small as possible in order to minimize memory consumption and to free up space for other tasks or processes.

The input or control variable is clearly the maximum cache size n. What is the output variable? Each request results in a hit or a miss, so the output—after each request—can be expressed as a Boolean variable. In order to arrive at the hit *rate*, we will need to calculate the "average" number of successes over the most recent k requests. This average rate will be the output variable.

This raises the question of how large k needs to be. The answer is unexpectedly large! Because each request results in either success or failure, requests can be regarded as Bernoulli trials. The uncertainty associated with estimating the success rate of Bernoulli trials[1] from a sample of size k is approximately $\pm 0.5 / \sqrt{k}$. If we want to control the hit rate to within a few percent, we need to know the actual success rate to within about ± 0.05 or better. Therefore, we need a sample of size $k \approx 100$. This may come as a bit of a surprise—with a desired hit rate of 0.9, would not a sample of 10 to 20 requests be sufficient? No, because the variations in the hit rate estimate would be too large to allow for meaningful control. This requirement on the sample size constitutes a fundamental limitation of the system: it has nothing to do with the chosen control architecture but is simply a consequence of the stochastic nature of the system and the central limit theorem.

We now have a choice between treating the "smoothing filter" (used to calculate the trailing average) as a separate component within the control loop, or instead as part of the controlled plant. In the former case, we would insert the filter into the return path of the control loop; in the latter case, the filter would be considered part of the cache itself, completely hiding the Boolean hit/miss signal from our view. (See Figure 13-1.) We need to remember, however, that this filter—with its relatively "long" memory— will alter the dynamics of the controlled system. Precisely for this reason it is convenient to treat the filter as part of the controlled plant: in that way, all the nontrivial dynamics are bundled into a single component.

---

1. For an in-depth discussion, see *An Introduction to Probability Theory and Its Applications, Volume 1* by William Feller (1968).

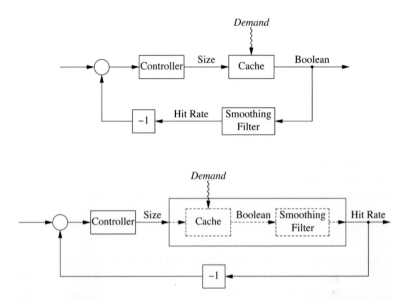

Figure 13-1. Two possible ways to view the loop architecture: we can treat the smoothing filter either as a separate component inserted into the return path (top) or as part of the plant itself (bottom). For our purposes, the latter architecture is more useful.

Another subtlety concerns the way we treat time. There are two basic options: we can treat cache requests and control actions as either synchronous or asynchronous (see Figure 13-2). In the former case, each request leads to a new output value (either hit or miss, leading to a new value of the smoothed hit rate). It then makes sense to require that a new input value (that is, a new maximum cache size) can be specified *only* in response to an output value; control actions are therefore synchronized with cache requests. Alternatively, we can assume that cache requests occur completely independently from control actions, with requests and control inputs each being processed in their own "thread," so to speak. Both designs are possible, but each is based on a different notion of "time." In the synchronous case, time progresses not by itself but only in steps defined by the occurrence of cache requests; we can call this "event time." In the asynchronous case, though, time progresses normally as "wall-clock time." In the example code at the end of the chapter, we will implement the synchronous, event-time design (mostly for demonstration purposes, since other case studies are more naturally expressed in terms of wall-clock time).

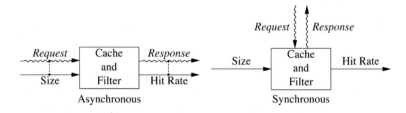

Figure 13-2. Schematic representation of the two ways that control actions and cache requests can relate to each other: in the synchronous case (left), control actions occur only in conjunction with cache requests; in the asynchronous case (right), the flow of control actions is completely independent from the stream of cache requests. We will be using the model on the left.

Finally, a word on the "demand," by which I mean the stream of cache requests. This demand is entirely outside our control—it just happens! But it clearly affects the controlled output variable. For instance, if the selection of most frequently accessed items changes, then we should expect a drop in hit rate until the cache has been repopulated. If the range of selected items changes, the hit rate will suffer until the cache has been resized. The access pattern is therefore part of the dynamics of the cache but in a way that is not directly visible to us; we will need to treat it in a black-box manner. Such disturbances occurring within the controlled system are referred to as "load disturbances" (because they arise in the "load," which is the system that is driven by the controller). For the caching system, where the setpoint is held constant over long stretches of time, a controller will be evaluated by how well it is able to cope with these load disturbances.

Cache Misses as Manufacturing Defects

The entire discussion so far has been in terms of cache hits and misses and the desire to maintain a specific hit rate. If we think of the cache as a manufacturing plant, then we can identify cache misses with defective items leaving the production line. Most of the discussion—in particular, the need for smoothing (and the size of the required smoothing filter) as well as the choice between synchronous and asynchronous modes of operation—carry over to any form of statistical process control. In a process control situation, we want to maintain some defect rate among discrete items leaving a manufacturing line. Whenever there is an input variable that controls the operation of the production line, then it is possible to use a feedback loop similar to the cache control loop introduced here. For example, we can imagine a production line turning out semiconductor components that need to undergo "high-temperature tempering" as part of their manufacturing process, where higher temperatures result in fewer defects. To conserve energy, we want to use the lowest possible temperature while still maintaining an acceptable defect rate. This can be accomplished with a feedback loop like the one described in this chapter. We can easily think of further examples. The setting changes, but the concepts stay the same.

Measuring System Characteristics

To begin with, we gather information about the behavior of our system, or "plant," as described in Chapter 8. As a first step we measure the static "process characteristic." In the present case, this is the relation between the cache size and the resulting hit rate. We will assume that the pattern of requests changes over the course of the day, so the graph (see Figure 13-3) shows data taken at different times.[2] From the figure we can already get a rough idea for the typical cache size and for the magnitude of the required changes in size.

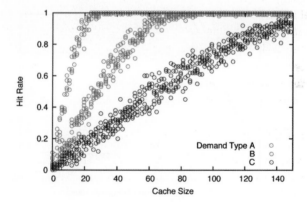

Figure 13-3. Measuring the cache's static process characteristic. The graph shows the steady-state hit rate, for three different traffic patterns, as a function of the cache size.

2. The data was collected from a simulated system. We will show and discuss the simulation code later in this chapter.

As a second step, we conduct a step test to measure the dynamic response or "process signature." Initially, the cache is empty and is limited to size 0. We then suddenly increase its size to some value n_0 and wait until the hit rate has stabilized. Now we can estimate the time constants of the process using one of the methods described in Chapter 8. From the process characteristic (Figure 13-3) we can see that a cache size of $n_0 = 40, ..., 70$ is about right if we want to attain a hit rate of 0.7. (Let's suppose that rate is our desired operating point and we mostly expect demand of type B.) After applying an input change of size $n_0 = 40$, we observe the hit rate. The results are shown in Figure 13-4 together with a Ziegler–Nichols construction and the "best fit" of the simple model $K(1 - \exp(-(t - \tau)/T))$. The parameter values obtained with the Ziegler–Nichols method are

$$\lambda = \frac{12}{90} \qquad \tau = 12$$

and

$$K = 0.71 \qquad \tau = 9.5 \qquad T = 54.1$$

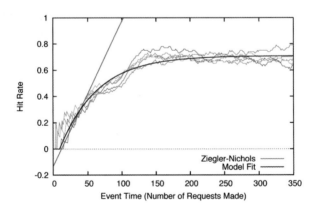

Figure 13-4. Measuring the cache's dynamic process reaction curve. The graph shows the evolution of the hit rate in response to a sudden change in cache size from n = 0 to n = 40, which occurred at time t = 0. Data from five simulation runs is shown together with a Ziegler–Nichols construction and an analytical model.

At this point, we need to take a brief detour. The system in question is stochastic in nature: we don't know what items will be requested next, and if we run the experiment again then a different set of items will be requested—and in a different order. Hence, each curve shown in Figure 13-4 is only one specific *example* from the population of all possible response curves. We should therefore evaluate the importance of fluctuations between different trials. The figure shows results from five separate runs, and it is clear that the variations between different trials are comparable to the inaccuracy of the model itself. The model has been "fitted" using the combination of results from all runs (after discarding the first 20 events from each run).

Controller Tuning

We can now use the experimental parameter values together with one of the available tuning methods to determine the controller gains for a PI controller. Table 13-1 reports results for the three methods discussed in Chapter 9: the simple Ziegler–Nichols method, as well as the Cohen–Coon and the AMIGO methods. Not unexpectedly, all three methods give results that are roughly comparable but are not identical; they indicate the range of values over which we should expect to find acceptable behavior. (When using the formulas, bear in mind that the observed gain λ or K must be divided by the magnitude of the step input change, $n_0 = 40$. Thus, the value to use in the Cohen–Coon and AMIGO formulas is $K/40 = 0.018$.)

Table 13-1. Values for the gain parameters as suggested by the different tuning methods.

	Ziegler–Nichols	Cohen–Coon	AMIGO
$k = k_p$	270	98.7	80.4
T_i	36	23.0	40.5
$k_i = k/T_i$	7.5	4.3	2.0

Figure 13-5 through Figure 13-8 show the results of several simulation runs using a variety of values for the controller gains. The simulation has been set up in such a way that a steplike load disturbance occurs twice during each simulation run. At time step $t = 3000$, the range of items being requested grows by roughly a factor of 2, at $t = 5000$, the range shrinks again but also moves (so that fewer but different items are being requested, requiring the cache to repopulate itself).

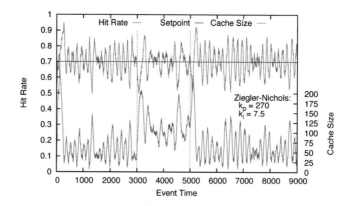

Figure 13-5. Simulation results for the cache size and cache hit rate when using the Ziegler–Nichols tuning method. The nature of the request traffic changes twice in the course of the observation period (for t = 3000 and t = 5000).

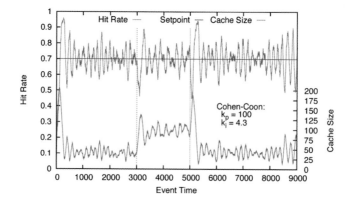

Figure 13-6. Simulation results for the cache size and cache hit rate when using the Cohen–Coon tuning method. (See Figure 13-5 for further details.)

Figure 13-5 shows a simulation run using the Ziegler–Nichols values for the controller gains. For the present system, the Ziegler–Nichols method results in rather poor behavior: both hit rate and cache (control input and process output) exhibit large-amplitude oscillations. In response to a load change, the system overshoots significantly and takes some time to settle again.

Figure 13-6 shows data obtained in a simulation based on the Cohen–Coon values for the parameter gains. The amplitude of the control input (cache size) oscillations is reduced, and the response to load disturbances is better. The hit rate, however, does not track the setpoint particularly faithfully, oscillating constantly and with an amplitude that is about 10 percent of the average value.

In Figure 13-7 we see the results obtained using the AMIGO method. The behavior is now quite good—in particular, the system responds to load changes quickly and without overshooting. The oscillatory behavior in the steady state is much reduced; the remaining wiggles are mostly due to random noise.

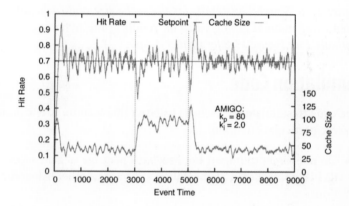

Figure 13-7. Simulation results for the cache size and cache hit rate when using the AMIGO tuning method. (See Figure 13-5 for further details.)

Finally, in Figure 13-8 we see an attempt to improve on the AMIGO method by manually adjusting the controller gains. Since both the Ziegler– Nichols and the Cohen–Coon method lead to much larger values for the proportional term, we now try increasing k_p again while keeping k_i at its AMIGO value. The greater proportional gain leads to a slightly faster response overall, allowing the system to respond more quickly to momentary tracking errors. The result is a further reduction in the amplitude of the output noise, although it leads to larger overshoot in response to load changes.

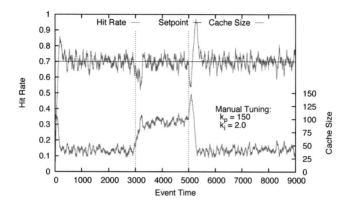

Figure 13-8. Simulation results for the cache size and cache hit rate when using manual tuning based on results of the systematic tuning methods. (See Figure 13-5 for further details.)

Simulation Code

The code to simulate the cache is listed in this section. A few points deserve particular attention:

- Do not forget to import the feedback package to gain access to its PID controller implementation and various other helper functions.

- The Cache constructor takes a reference to a function modeling the "demand" of item requests. This function takes the current time step as its argument (this way, one can model demand that changes over time), and it should return a value specifying the requested object (the object key). Since the implementation of the cache is built as a Python dictionary, the object key must be usable as a key in a Python dictionary. Typically, the demand function will wrap some random-number generator to simulate item requests.

- As discussed earlier in this chapter, in the current implementation, all control actions are synchronized with object requests: whenever the cache's work() function is called (signifying a control action), a new object request is generated by calling the demand function.

- The bulk of the work() function is an implementation of the actual cache protocol. When an item needs to be removed from the cache to make room for a new one, the item that has not been accessed for the longest time is removed before the new item is inserted.

```python
import feedback as fb

class Cache( fb.Component ):
    def __init__( self, size, demand ):
        self.t = 0           # internal time counter,
                             #   needed for last access time

        self.size = size  # size limit of cache
        self.cache = {}   # actual cache:
                          #   cache[key] = last_accessed_time

        self.demand = demand # demand function

    def work( self, u ):
        self.t += 1

        self.size = max( 0, int(u) ) # non-negative integer

        i = self.demand( self.t )   # the "requested item"

        if i in self.cache:
            self.cache[i] = self.t   # update last access time
            return 1

        if len(self.cache) >= self.size: # must make room
            # number of elements to delete:
            m = 1 + len(self.cache) - self.size

            tmp = {}
            for k in self.cache.keys():
                # key by last_access_time:
                tmp[ self.cache[k] ] = k

            for t in sorted( tmp.keys() ):
                # delete the oldest elements:
                del self.cache[ tmp[t] ]
                m -= 1
                if m == 0:
                    break

        self.cache[i] = self.t       # insert into cache
        return 0
```

The work() function in the simple Cache implementation returns 0 or 1 to indicate whether the most recent request resulted in a cache miss or cache hit, respectively. The SmoothedCache implementation subclasses the simple implementation, but it also contains a filter to turn the binary outcomes into a continuous hit rate. Its constructor takes an additional argument avg that specifies the number of trailing requests that are averaged to form the hit rate.

```
class SmoothedCache( Cache ):
    def __init__( self, size, demand, avg ):
        Cache.__init__( self, size,  demand );
        self.f = fb.FixedFilter( avg )

    def work( self, u ):
        y = Cache.work( self, u )
        return self.f.work(y)
```

We also need to specify the demand function and a function to provide the current setpoint. Both of these functions take the current simulation time step as their sole argument.

```
def demand( t ):
    return int( random.gauss( 0, 15 ) )

def setpoint( t ):
    return 0.7
```

To model the load changes in the simulation runs for Figure 13-5 through Figure 13-8, I used the following demand() function, which changes the parameters used by the random-number generator, depending on the time step:

```
def demand( t ):
    if t < 3000:
        return int( random.gauss( 0, 15 ) )
    elif t < 5000:
        return int( random.gauss( 0, 35 ) )
    else:
        return int( random.gauss( 100, 15 ) )
```

With all these provisions in place, we can now instantiate a SmoothedCache as our "plant" and a PID controller and run them— for instance in a closed-loop arrangement—using one of the convenience functions from the feedback package. Since we are using event time, the width of the time step DT is set to 1 (counting the number of events that have occurred so far).

```
fb.DT = 1

p = SmoothedCache( 0, demand, 100 )
c = fb.PidController( 100, 250 )

fb.closed_loop( setpoint, c, p, 10000 )
```

CHAPTER 14

Case Study: Ad Delivery

The process of delivering advertisements on the Internet constitutes a possible application of feedback principles. It exhibits many of the characteristics that make feedback mechanisms desirable:

- The laws governing the process are either not known or only incompletely known; the process itself is opaque.
- The governing laws may change silently over time.
- The process is subject to random changes (because of fluctuations in web traffic).
- Yet, there is a clear goal; typically, this is the number of ad impressions to show every day or the amount of money to spend.

What is desired is a reliable mechanism for executing the plan in the face of uncertainty about the process and subject to the random fluctuations in web traffic.

The Situation

The specific situation we will consider consists of a publisher or advertising network that displays ads—ours as well as those from competing advertisers. We cannot directly control how often our ads are shown. Instead, we can name the maximum price that each showing (or impression) is worth to us, after which the publisher will make a selection based on this offer. All we know is that a higher offered price tends to result in a greater number of ad impressions.

The system we want to control is the publisher. The control input is the price, and the control output is the number of impressions served. Knowing that a higher price means more impressions provides the tiny but necessary bit of process knowledge required to set up a feedback scheme.

I assume that we can set our price once every 24 hours and that we will learn the outcome (that is, the number of impressions truly delivered for this price) by the end of that period. Of course, there is nothing magical about the 24-hour period—any other fixed time (such as 6 hours or 48 hours) would work as well. What *is* relevant is that we will not learn about the results of the most recent run until the moment we have to choose a new input. This kind of dynamics is typical of computer systems but is quite unlike the behavior of items in the physical world: instead of a gradual response to a control input, we get a steplike response that is complete but delayed by one time step.

The setpoint for this control problem consists of the number of impressions we wish to generate every day. Let's suppose that we already have a plan that spells this number out.

Measuring System Characteristics

There is very little that we know offhand about the process that we want to control except for the directionality of the input/output relationship (higher prices lead to a greater number of impressions). In the first step, we measure the static process characteristic (Chapter 8).

For several values of the input (the price), we observe the resulting output (the number of impressions). Since the number of impressions will fluctuate from day to day because of the randomness of web traffic, we should repeat every measurement several times. The results might look like Figure 14-1.[1] There is a minimum price that needs to be offered to receive any traffic; for higher prices, we see that the number of delivered impressions grows with the price but not linearly. The range of outcomes for each fixed price gives us an estimate for the amplitude of the day-to-day fluctuations. So far, so good.

Figure 14-1 looks simple, but it hides a great deal of *practical* difficulty. Recall that we must wait an entire day for each data point in the graph.

---

1. The data was collected from a simulated system. We will show and discuss the simulation code later in this chapter.

The figure shows 10 trials for each price, and a total of 100 different prices. If we conduct this experiment sequentially, it will take *three years* to collect all the data—by which time most of it will be obsolete. It is clear that most of the time we will have to content ourselves with much less complete test coverage—perhaps just two or three data points to find the order of magnitude of the quantities involved. (The inability to run exhaustive experiments is a recurrent issue when working with control systems.)

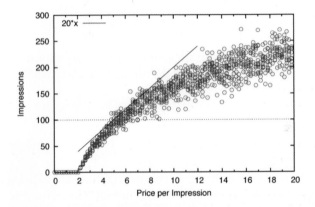

Figure 14-1. The static process characteristic: if we are willing to pay a certain price per impression, how many impressions do we obtain?

We do not need to conduct an experiment to find the *dynamic* response of the system because we already know it: full response, but one day later. This is enough to set up a basic control loop.

Establishing Control

To establish control, we set up a basic control loop for our system using a standard PID controller K. The controller converts the difference between the daily impression goal (the setpoint) and the number of impressions actually delivered on the previous day into a new price, which is then passed to the publisher (see Figure 14-2).

In order to make things concrete, we *must* choose numerical values for the gain parameters in the PID controller. Unfortunately, none of the tuning methods discussed in Chapter 9 seem to be of much help. These methods all assume that there is a measurable delay τ and a time constant T in order to arrive at numerical values that can be used in a control loop implementation. In the present situation, however, the response is *instantaneous*. What should we do?

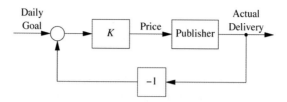

Figure 14-2. Loop architecture for the current case study.

The situation is actually much simpler than it may appear. Near the end of Chapter 9 we saw that all the tuning formulas could be written in the form

$$K(s) = A \left[a + \frac{b}{s} + cs \right] \frac{\Delta u}{\Delta y}$$

where Δu is the static change in control input needed to bring about a change Δy in the steady-state control output. The factor A is an adjustment to the static input/output relation that takes the system dynamics into account.

In the present situation, we are trying to control a system that has no (nontrivial) dynamics: it responds immediately. In other words, there is no need for the dynamic adjustment factor in the controller—we can basically ignore the A factor.

The static factor $\Delta u / \Delta y$ can be obtained directly from the process characteristic: Figure 14-1 tells us that if we start with a price of $5, then a change by $\Delta u = \pm\$1$ will result in a change in output of roughly $\Delta y = \pm 20$ impressions; therefore,

$$\frac{\Delta u}{\Delta y} = \frac{1}{20}$$

Finally, we will choose (somewhat arbitrarily)

$$a = 0.5 \qquad b = 0.25 \qquad c = 0.0$$

as values for the term-specific coefficients (inside the central brackets). The actual controller gains consist of the combination of this coefficient with the static gain:

$$k_p = a\frac{\Delta u}{\Delta y} \qquad k_i = b\frac{\Delta u}{\Delta y} \qquad k_d = c\frac{\Delta u}{\Delta y}$$

The behavior that is found when a controller with these specifications is placed into the loop is shown in Figure 14-3. As we can see, the performance is not great but is entirely acceptable.

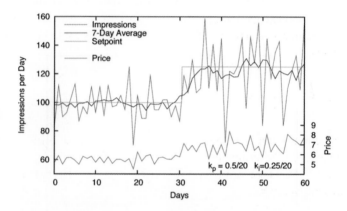

Figure 14-3. Performance of the closed loop when using the "default" controller gains.

Improving Performance

Looking at Figure 14-3, we can see that the system does track the setpoint and responds to the setpoint change reasonably quickly (within four to five days). But the daily fluctuations in process output are large (as much as ±30 percent). What is worse, the process *input* also fluctuates. If there is a fixed fee or administrative cost associated with each price change that we submit to the publisher, such noisy behavior is clearly undesirable.

In an attempt to make the system less noisy, we might try to reduce the proportional term and rely more on the slower-acting integral term. In fact, we can dispense with the proportional term entirely! As long as the setpoint does not change for several days at a time, the changes in the tracking error from one day to the next are due entirely to random fluctuations in the web traffic. In the present situation, fluctuations on consecutive days are statistically independent; hence there is no way to predict tomorrow's value from today's. So the proportional term, which acts on the current value of the tracking error, is useless: there is no way that the momentary value of the tracking error can be used to reduce the tracking error in the future.

Figure 14-4 and Figure 14-5 show the system under strictly integral control. In Figure 14-4, the integral gain is the same as in Figure 14-3 ($k_i = 0.25/20$) but the proportional gain is now zero ($k_p = 0$). Compared to the previous situation, the amplitude of the noise has been reduced and the price (the control input) is much steadier; however, the system is also more sluggish.

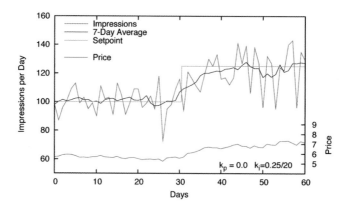

Figure 14-4. Performance of the closed loop under strictly integral control. For k_i = 0.25/20, the behavior is sluggish: it takes more than 10 days to adjust fully to the setpoint change.

Whereas before it reacted to a setpoint change within a few days, now it takes well over 10 days to fully adjust to a new setpoint value. This is a problem—advertising campaigns typically only run for weeks or months, so a delay that exceeds a week is probably not acceptable.

We might try improving the system's responsiveness by increasing the integral gain, but any increase in speed comes at the expense of increased instability: the system begins to oscillate wildly from one time step to the next (Figure 14-5). This is a common problem in control systems with discrete time steps and immediate-response dynamics. Because there is no partial response, the system easily falls into this particular type of rapid control oscillation if the corrective actions are too large. That tendency imposes relatively tight limits on the controller gains we can use in systems of this kind.

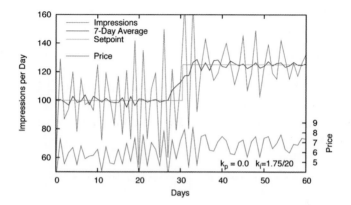

Figure 14-5. Performance of the closed loop under strictly integral control. Increasing the controller gain to $k_i = 1.75/20$ leads to control oscillations.

These observations already suggest the solution: we need to introduce a filter that will slow things down and "round off" the responses. This results in a loop architecture like the one shown in Figure 14-6 (bottom). A recursive filter

$$z_t = ay_t + (1 - \alpha)z_{t-1} \qquad \alpha = 0.125$$

is introduced into the return path. The filter output lags behind but is also smooth, so it makes sense to use proportional control again in order to speed up the response. With $k_p = 1.0/20$ and $k_i = 0.125/20$, the observed behavior seems to strike a good balance (Figure 14-6, top). Notice in particular how little the price changes over time—except for the sudden (and properly sized) jump in response to the setpoint change.

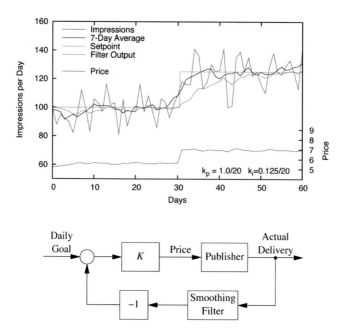

Figure 14-6. The effect of including a smoothing filter into the loop. Now that the signal is smooth, it makes sense to include proportional control again.

Variations

The solution just presented seems satisfactory, but there are additional variations and enhancements that we don't explore in detail in this case study. Some directions are summarized next.

Cumulative Goal

In the preceding solution, we used the *daily* impression target as setpoint. We can slice the problem differently by using the *cumulative* impression target as setpoint. From a business perspective, this is arguably a more meaningful design because the quantity that we ultimately want to control is usually the number of impressions served over the entire duration of a campaign—the daily target is merely derived from that.

In order to use the cumulative target, we must insert an additional integrator into the return path of the control loop (Figure 14-7, bottom). The process continues to report the number of impressions served per day and the integrator adds them up, reporting the cumulative impressions served, so that their value can be compared to the cumulative goal. Of course, the setpoint must now steadily *increase* in value. The resulting behavior is shown in Figure 14-7 (top). (For graphing purposes, the daily setpoint value is shown.) We should keep in mind, however, that this solution has few (if any) advantages over the one using a daily goal and a smoothing filter. The integrator just introduced merely replicates the behavior of the controller's integral term in the previous solution.

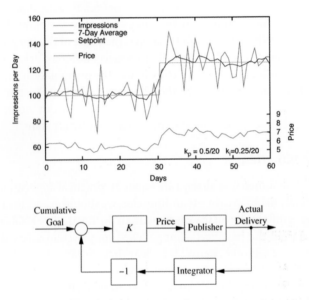

Figure 14-7. Including an integrator on the return path while using the cumulative instead of the daily goal. In this scenario, the setpoint is not constant but instead is a ramp input.

Gain Scheduling

In our discussions so far the target was always in the range of 100–125 impressions per day, requiring a price between $5.50 and $7.00. As can be seen from the static process characteristic in Figure 14-1, a static controller gain factor of $\Delta u/\Delta y = 1/20$ is appropriate for this operating point. Yet if we want to use a very different goal, such as 200–250 impressions per day, then a larger gain factor (closer to $\Delta u/\Delta y = 1/7.5$) would be required because the system becomes less price sensitive at higher traffic volumes. If we would build a controller that chooses the appropriate gain factor from a lookup table (based on the setpoint value), that would be an example of *gain scheduling* (see Chapter 11).

Integrator Preloading

The dynamic response to the setpoint change of the final system (Figure 14-6) is quite satisfactory. Nevertheless, it may be desirable to speed up the response to a setpoint change that is known (in advance) to occur. One way to do so is by *integrator preloading*. To achieve the daily delivery goal, the controller must produce a nonzero value for the price that is fed to the publisher. In the steady state, this constant offset is produced entirely by the integral term in the controller (see Chapter 4). When the setpoint changes, the value of the integral term needs to change as well in order to produce a different price. We can use this knowledge and adjust the value of the integral term inside the controller concurrently with the change in setpoint. In particular, when the system is first started up, this approach can significantly reduce the time it takes the system to reach its steady state.

Weekend Effects

So far we assumed that all days are equal. In practical applications this is not likely to be the case—if nothing else, we should expect the system to behave differently on the weekend than during the week. Given that it takes even the best-performing closed-loop system a few days to respond to a setpoint change, it is clear that we cannot simply ignore weekends. A practical solution would be to have two entirely different controller instances for weekday and weekend traffic. These instances would retain their memory (in the form of the value of their integral terms) between invocations, but signals would be directed to the correct instance only at any given moment.

Simulation Code

The model for the publisher that was used in the simulations is given next. Its primary responsibility is to create a value for the "impressions delivered," given a "price." For the purpose of this simulation, I create the number of impressions as random numbers drawn from a Gaussian distribution. The mean of the distribution depends logarithmically on the price. (The latter choice is common in decision-theoretic modeling: the "value" of an item increases only logarithmically with the price, as when a car that is 10 times more expensive can only drive twice as fast.) Please keep in mind that this implementation is only a *model* that has been chosen to demonstrate the basic process. Other choices could (and possibly should) be made.

```
include math
include random
include feedback as fb

class AdPublisher( fb.Component ):

    def __init__( self, scale, min_price, relative_width=0.1 ):
        self.scale = scale
        self.min = min_price
        self.width = relative_width

    def work( self, u ):
        if u <= self.min:      # Price below min: no impressions
            return 0

        # "demand" is the number of impressions served per day
        # The demand is modeled (!) as Gaussian distribution
        # with a mean that depends logarithmically on the
        # price u.

        mean = self.scale*math.log( u/self.min )
        demand = int( random.gauss( mean, self.width*mean ) )

        return max( 0, demand ) # Impression demand is greater
                                # than zero.
```

We can use this definition of the AdPublisher to form a control loop, using a PID controller and one of the convenience functions from the simulation framework:

```
def closedloop( kp, ki, f=fb.Identity() ):
    def setpoint( t ):
        if t > 1000:
            return 125
        return 100

    k = 1.0/20.0

    p = AdPublisher( 100, 2 )
    c = fb.PidController( k*kp, k*ki )

    fb.closed_loop( setpoint, c, p, returnfilter=f )
```

We need to select the length of the time step before we can run the simulation. We perform one control action per day. If we measure time in days, then the length of each step is equal to 1.

Running the closed loop arrangement, including the smoothing filter, can then be accomplished using the following call:

```
fb.DT = 1

closedloop( 1.0, 0.125, fb.RecursiveFilter(0.125) )
```

CHAPTER 15

Case Study: Scaling Server Instances

Consider a data center. How many server instances do you need to spin up? Just enough to handle incoming requests, right? But precisely how many instances will be "enough"? And what if the traffic intensity changes? Especially in a "cloud"-like deployment situation—where resources can come and go and we only pay for the resources actually committed—it makes sense to exercise control constantly and automatically.

The Situation

The situation sketched in the paragraph above is common enough, but it can describe quite a variety of circumstances, depending on the specifics. Details matter! For now, we will assume the following.

- Control action is applied periodically—say, once every second, on the second.

- In the interval between successive control actions, requests come in and are handled by the servers. If there aren't enough servers, then some of the requests will not be answered ("failed requests").

- Requests are not queued. Any request that is not immediately handled by a server is "dropped." There is no accumulation of pending requests.

- The number of incoming and answered requests for each interval is available and can be obtained.

- The number of requests that arrive during each interval is (of course) a random quantity, as is the number of requests that a server does handle. In addition to the second-to-second variation in the number of incoming requests, we also expect slow "drifts" of traffic intensity. These drifts typically take place over the course of minutes or hours.

This set of conditions defines a specific, concrete control problem. From the control perspective, the absence of any accumulation of pending requests (no queueing) is particularly important. Recall that a queue constitutes a form of "memory" and that memory leads to more complicated dynamics. In this sense, the case described here is relatively benign.

The description of the available information already implies a choice of output variable: we will use the "success ratio," which is the ratio of answered to incoming requests. And naturally we will want this quantity to be large (that is, to approach a success rate of 100 percent). The number of active servers will be the control input variable.

Measuring and Tuning

Obtaining the static process characteristic is straightforward, and the results are not surprising (see Figure 15-1): the success rate with which requests are handled is proportional to the number of servers.[1] Eventually, it saturates when there are enough servers online to handle all requests. The slope and the saturation point both depend on the traffic intensity.

In regards to the dynamic process reaction curve, we find that the system responds without exhibiting a partial response. If we request n servers then we will get n servers, possibly after some delay. But we will *not* find a partial response (such as $n/4$ servers at first, $n/2$ a little later, and so on). This is in contrast to the behavior of a heated pot: if we turn on the heat, the temperature *begins* to increase immediately yet reaches its final value only gradually. Computer systems are different.

We can obtain values for the controller parameters using a method similar to the one used in Chapter 14: the tuning formulas (Chap-

---

1. The data was collected from a simulated system. We will show and discuss the simulation code later in this chapter.

ter 9) consist of the static gain factor $\Delta u / \Delta y$ multiplied by a term that depends on the system's dynamics. Because in the present case the system has no nontrivial dynamics, we are left with the gain factor. From the process characteristic (Figure 15-1), we find that

$$\frac{\Delta u}{\Delta y} \approx \frac{10}{1}$$

for intermediate traffic intensity. Taking into account the numerical coefficients for each term in the PID controller, we may choose

$$k_p = 1 \qquad k_i = 5$$

as controller gains. Figure 15-2 shows the resulting behavior for setpoint values much less than 100 percent. The system responds nicely to changes in both setpoint and load.

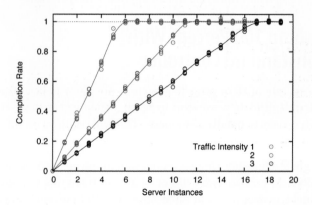

Figure 15-1. The steady-state relationship between the completion rate and the number of server instances (the static process characteristic).

The system is very oscillatory, but in this case this is not due to a badly tuned controller, but to the inability of the system to respond accurately to its control input. This is especially evident in the middle section of the right-hand graph in Figure 15-2 (between time steps 100 and 200): six server instances are clearly not enough to maintain the desired completion rate of 0.6, but seven server instances are too many.

Because we can't have "half a server," the system will oscillate between the two most suitable values.

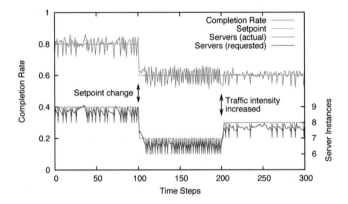

Figure 15-2. Closed-loop behavior when using a PID controller and desired completion rates that do not approach 100 percent.

Reaching 100 Percent With a Nonstandard Controller

A success rate of 80 or even 90 percent is not likely to be considered sufficient—ultimately, we want to reach a state where there are just enough servers to handle *all* requests. This will require a modification of the usual approach as well as a nonstandard (nonlinear) controller.

We begin by choosing a high setpoint, something on the order of 0.999 or 0.9995. In practice, we will not be able to track this setpoint very accurately. Instead, we will often observe a 100 percent completion rate simply because the number of servers sufficient to answer 999 out of 1,000 requests has enough spare capacity to handle one more. This is a consequence of the fact that we can add or remove only entire server instances, and would occur even if there was no noise in the system.

It is important that the setpoint is *not* equal to 1.00. The goal of achieving 100 percent completion rate is natural, but it undermines the feedback principle. For feedback to work, it is necessary that the tracking error can be positive *and* negative. With a setpoint value of 1.00, the tracking error cannot be negative:

tracking error e = setpoint r – process output y

and this removes the pressure on the system to adapt. (After all, it is easy enough to achieve a 100 percent completion rate: just keep adding servers. But that's not the behavior we want.)

As long as the setpoint is less than 1.00, an actual completion rate of 100 percent results in a finite, negative tracking error. This error accumulates in the integral term of the controller. Over time, the integral term will be sufficiently negative to make the system *reduce* the number of instances. This forces the system to "try out" what it would be like to run with fewer instances. If the smaller number of servers is not sufficient, then the completion rate will fall below the setpoint and the number of servers will increase again. But if the completion rate can be maintained even with the lower number of servers (because the number of incoming requests has shrunk, for instance), then we will have been able to reduce the number of active servers.

But now we face an unusual asymmetry because the tracking error can become much more positive than it can become negative. The actual completion rate can never be larger than 1.0, and with a setpoint of (say) $r = 0.999$, this limits the tracking error on the negative side to $0.999 - 1.0 = -0.001$. On the positive side, in contrast, the tracking error can become as large as 0.999, which is more than two orders of magnitude larger. Since control actions in a PID controller are proportional to the error, this implies that control actions that tend to increase the number of servers are usually two orders of magnitude greater than those that try to decrease it. That's clearly not desirable.

We can attempt various ad hoc schemes in order to make a PID-style controller work for this system. For instance, we could use different gains for positive and negative errors, thus compensating for the different range in error values (see Figure 15-3). But it is better to make a clear break and to recognize that this situation calls for a different control strategy.

Fortunately, the discussion so far already contains all the needed ingredients—we just need to put them together properly. Two crucial observations are:

- The *magnitude* of the error is not very important in the present situation; what matters is its *sign*.

- We are not able to adjust the control input continuously; the number of servers must always be a whole (positive) integer.

These two observations suggest the following simple control strategy:

1. Choose $r = 1.0$ as setpoint. (100 percent completion rate—this implies that the tracking error can never be negative.)
2. Whenever the tracking error $e = r - y$ is positive, increase the number of active servers by 1.
3. Do nothing when the tracking error is zero.
4. Periodically *decrease* the number of servers by 1 to see whether a smaller number of servers is sufficient.

Step 4 is crucial: it is only through this periodic reduction in the number of servers that the system responds to decreases in request traffic. We can make the controller even more responsive by scheduling trial steps more frequently after a decrease in server instances than after an increase.

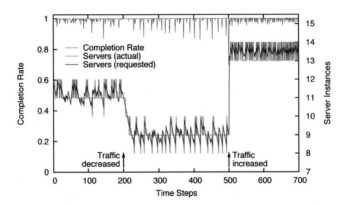

Figure 15-3. Choosing a high setpoint (such as r = 0.9995) is a problem because the tracking error can become much more positive than it can become negative. Here we use a controller that uses different gains for negative and positive tracking errors.

Figure 15-4 shows typical results, including the response to both a decrease and an increase in incoming traffic. The operation is very stable and adjusts quickly to changes. Another benefit (compared to the behavior shown in Figure 15-3) is that the system oscillates far less.

The number of server instances is constant, except for the periodically scheduled "tests."

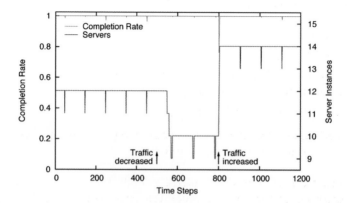

Figure 15-4. Much better results can be obtained using a special controller that increases or decreases the number of server instances based only on the sign, not the magnitude, of the tracking error. (See main text for details.)

Dealing with Latency

Until now we have assumed that any new server instances would be available immediately, one time step after the request. In practice, this is not likely to be the case: spinning up new instances usually takes a while, maybe a couple of minutes. Because the interval between control actions is on the order of seconds, this constitutes a delay of two orders of magnitude! That is too much time to ignore.

When it comes to latency, textbooks on control theory usually recommend that one "redesign the system to avoid the delays." This may seem like unhelpful advice, but it is worth taking seriously. It may be worth spending significant effort on a redesign, because the alternatives (such as the "Smith predictor"; see Chapter 11) can easily be even more complicated.

In the present situation, one way to deal with the latency issue is the introduction of a set of "warm standbys" together with a second feedback loop that controls them (see Figure 15-5). There must already be a network switch or router present, which serves as a load balancer and distributes incoming requests to the various server instances. This switch should be able to add or remove connections to server instances very quickly. As long as we always have a sufficient number of spun-up yet inactive server instances standing by, we can fulfill requests for additional resources as quickly as the router can open those connections (quite possibly within one or two time steps).

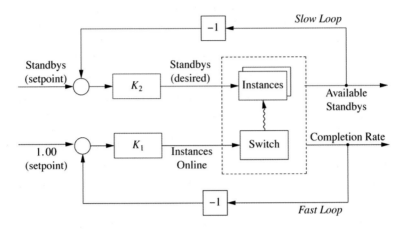

Figure 15-5. A possible control architecture to deal with latency in spinning up additional server instances. A "fast" loop controls the number of instances currently active, based on the rate at which requests are completed; a "slow" loop maintains a pool of warm standbys.

In addition, we will need a *second* control loop that regulates the number of available standbys. We must decide how many standbys we will need in order to maintain our desired quality of service in the face of changing traffic patterns: that will be the setpoint. The controller in this loop simply commissions or decommissions server instances in order to maintain the desired number of reserves. This second loop will act more slowly than the first; its primary time scale is the time it takes to spin up a new instance. In fact, we can even set the sampling interval for this second loop to be equal to the time it takes to spin up a new server. With this choice of sampling interval, new instances are again available "immediately"—the delay has disappeared. (There is no benefit in sampling faster than the system can respond.)

Simulation Code

The system that we want to simulate is a server farm responding to incoming requests. We must decide how we want to model the incoming requests. One way would be to create an event-based simulation in which individual requests are scheduled to occur at specific, random moments in time. The simulation then proceeds by always picking the next scheduled event from the queue and handling it. In this programming model, control actions need to be scheduled as events and placed onto the queue in the same way as requests.

We will use a simpler method. Instead of handling individual requests, we model incoming traffic simply as an aggregate "load" that arrives between control actions. Each server instance, in turn, does some random amount of "work," thereby reducing the "load." If the entire "load" is consumed in this fashion, then the completion rate is 100 percent; otherwise, the completion rate is the ratio calculated as the completed "work" divided by the original "load."

The implementation is split into two classes, an abstract base class and the actual ServerPool implementation class. (This design allows us to reuse the base class in Chapter 16.)

```python
import math
import random
import feedback as fb

class AbstractServerPool( fb.Component ):
    def __init__( self, n, server, load ):
        self.n = n              # number of server instances
        self.queue = 0          # number of items in queue

        self.server = server    # server work function
        self.load = load        # queue-loading work function


    def work( self, u ):
        self.n = max(0, int(round(u))) # server count:
                                       # non-negative integer

        completed = 0
        for _ in range(self.n):
            completed += self.server() # each server does some
                                       # amount of work

            if completed >= self.queue:
                completed = self.queue # "trim" to queue length
                break                  # stop if queue is empty
```

```
        self.queue -= completed        # reduce queue by work
                                        # completed

        return completed


    def monitoring( self ):
        return "%d %d" % ( self.n, self.queue )


class ServerPool( AbstractServerPool ):
    def work( self, u ):
        load = self.load()        # additions to the queue
        self.queue = load         # new load replaces old load

        if load == 0: return 1    # no work:
                                  # 100 percent completion rate

        completed = AbstractServerPool.work( self, u )

        return completed/load     # completion rate
```

When instantiating a `ServerPool`, we need to specify two functions.
The first is called once per time step to place a "load" onto the queue;
the second function is called once for each server instance to remove
the amount of "work" that this server instance has performed. Both
the load and the work are modeled as random quantities, the load as
a Gaussian, and the work as a Beta variate. (The reason for the latter
choice is that it is guaranteed to be positive yet bounded: every server
does *some* amount of work, but the amount that each can do is finite.
Of course, other choices—such as a uniform distribution—are entirely
possible.) The default implementations for these functions are:

```
def load_queue():
    return random.gauss( 1000, 5 )        # default implementation

def consume_queue():
    # For the beta distribution : mean: a/(a+b); var: ~b/a^2
    a, b = 20, 2
    return 100*random.betavariate( a, b )
```

Figure 15-4 shows the behavior that is obtained using a special controller. This controller is intended to be used with a setpoint of, so that
the tracking error (which is also the controller input) can never be
negative. When the input to the controller is positive, the controller
returns, thus increasing the number of server instances by 1. When

---

the input is zero, the controller *periodically* returns −1, thereby reducing the number of server instances.

The controller uses different periods depending on whether the previous action was an increase or a decrease in the number of servers. The reason is that any increase in server instances is always driven by a positive tracking error and can therefore happen over consecutive time steps. However, a decrease in servers is possible only through periodic reductions—using a shorter period after a preceding reduction makes it possible to react to large reductions in request traffic more quickly.

```
class SpecialController( fb.Component ):
    def __init__( self, period1, period2 ):
        self.period1 = period1
        self.period2 = period2
        self.t = 0

    def work( self, u ):
        if u > 0:
            self.t = self.period1
            return +1

        self.t -= 1        # At this point: u <= 0 guaranteed!

        if self.t == 0:
            self.t = self.period2
            return -1

        return 0
```

The controller is an *incremental* controller, which means that it does not keep track of the number of server instances currently online. As an incremental controller, it computes and returns only the *change* in the number of server instances. We therefore need to place an integrator between the controller and the actual plant in order to "remember" the current number of server instances and to adjust it based on the instructions coming from the controller:

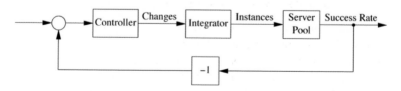

To simulate a closed loop using this controller is now simple:

```python
def setpoint( t ):
    return 1.0

fb.DT = 1

p = ServerPool( 0, consume_queue, load_queue )
c = SpecialController( 100, 10 )
fb.closed_loop( setpoint, c, p, actuator=fb.Integrator() )
```

CHAPTER 16

Case Study: Waiting-Queue Control

Queueing systems are ubiquitous, and they are always somewhat nerve-racking because they are inherently *unstable*. If we don't do something to handle the work items that are being added all the time, the queue will "blow up." Queueing systems *require* control to function properly. However, the very nature of queueing systems prevents the straightforward application of feedback principles.

On the Nature of Queues and Buffers

Why do queues exist? They do not simply occur because resources are insufficient to handle incoming requests. Insufficient resources are the reason for *exploding* queues but not for queues in general.

Queues exist to smooth out variations. If customers arrived at (say) a bank in regular intervals and if each transaction took the same amount of time, then there would be no queues: we could schedule tellers to meet demand exactly. In fact, many automated manufacturing lines work in precisely this way. But if customers arrive randomly and if transactions can take up different amounts of time, then there will be moments when demand can (temporarily) not be met. This happens even if the tellers' processing capacity is not less than the average arrival rate. There are only two ways to avoid such random accumulation of pending requests: scheduling arrivals periodically (thus reducing variation) or having an excess capacity of resources to handle requests (enough tellers to deal with anything). The first of these options is often not feasible, and the second is too expensive. Hence the need for queues and buffers.

What does this mean for queues from a control perspective? The most important point here is that it makes no sense to fix the queue length to one particular value: this would defeat the purpose of having a buffer in the first place! Instead, we usually want to restrict the length of the queue to some *range*, taking action (possibly drastic action) only if this range is violated. Keep in mind that it is often just as desirable to avoid an underflow of a buffer as it is to avoid an overflow: an empty buffer always means wasted resources (for example, tellers being idle).

We also need to ask what the overall goal of our control strategy is. In a physical environment, where a "buffer" is an actual device with a finite holding capacity (like a tank or an accumulating conveyor), we will mostly be concerned about avoiding overflows. But in more general situations, our goal will be to meet some quality-of-service expectation—and the length of the queue is not necessarily a good measure for the quality of service: a long queue that moves fast is more desirable than a short one that is stuck. In fact, the *waiting time* (or response time) will be a far better measure for our customer satisfaction. But the problem is a practical one: information about the waiting time may not be available! In order to base a system on waiting times, we must know the arrival time of each request, and this information is often not recorded. In contrast, the number of items pending is almost universally known. For this reason, it is desirable to have a strategy that works with only the length of the queue and without recourse to the actual waiting time.

Finally, a general concern when attempting to control queueing systems is that queues are *slow* and have difficult dynamics. If we only take action once a queue has become too long, it is—quite literally— too late. Not only has the queue already drifted from its target length, but we must now also process the accumulated backlog to bring the system back in line. The backlog requires additional resources to work through, which need to be decommissioned *before* the queue reaches its target length to prevent them from exhausting all the work and thereby underflowing the buffer. Buffers and queues are mechanisms for "memory" (in the sense of Chapter 3) and are hard to control. Therefore, early detection of change is essential. Note that this last requirement is in direct contradiction to the notion that we started with—namely, to let the length of a queue float!

The Architecture

We will solve the challenges posed by the waiting queue problem by using a *nested* (or "cascaded") control loop architecture (see Chapter 11). A fast-acting inner loop controls the actual plant, but the setpoint for the inner loop is provided by a controller in an outer loop.

Instead of controlling the length of the queue directly, in the present case the inner loop manages the *rate of change* of the length of the queue, adding or removing resources (servers) in order to keep the net change at the specified setpoint. The external loop, in contrast, will hold the overall length of the queue at the desired value without overwhelming the constraints of the buffer and without starving the downstream process. (See Figure 16-1.) The external loop provides the desired rate of change as the inner loop's setpoint.

The internal loop will act quickly because it is monitoring the *rate* at which the queue length grows or shrinks. As soon as the queue begins to increase, the inner loop will add further servers to consume work items (and vice versa). The external loop will act more slowly, which is fine, since the setpoint for the external loop (namely, the desired length of the queue) will be changed only rarely.

Setup and Tuning

To the external loop, the entire inner loop (shown by the dashed box in Figure 16-1) appears as a single component with exactly one input and output—in fact, the inner loop is the "plant," which is controlled by the outer loop. Nevertheless, there are two controllers involved, both of which need to be tuned.

Consider the inner loop first. The actual plant here is quite similar to the server farm from Chapter 15. Again, the control input is the number of server instances that are active and ready to handle incoming requests. The difference now is that incoming requests that cannot be handled are queued instead of being discarded.

The inner controller is tuned like other systems with immediate-response dynamics (see Chapter 9). We know that each server can handle about 100 requests per time step, so the static gain factor is

$$\frac{\Delta u}{\Delta y} = \frac{1}{100}$$

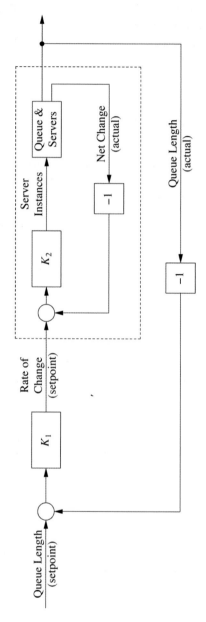

Figure 16-1. A nested (cascaded) loop architecture. The outer loop provides the setpoint for the fast-acting inner loop.

Because there are no nontrivial dynamics, we can ignore the dynamic adjustments to the controller gain; however, we must still choose the contributions for each of the individual terms in the controller (see Chapter 14). Let's use a PI controller with the following gains (including the static gain factor):

$$k_p^{inner} = \frac{0.5}{100} \qquad k_i^{inner} = \frac{0.25}{100}$$

(The superscripts on the gain factors emphasize that these are the gains for the controller K_2 of the inner loop in Figure 16-1. Gains for the external controller K_1 will not be specially marked.)

From this point on, the inner loop is simply treated as a black box: a component with a control input and output and with its own dynamics. As discussed in Chapter 8, we perform a step test on the inner loop to acquire the information needed to tune the *external* controller.

The results of the step test are shown in Figure 16-2.[1] Since the inner loop is a queue, we should not be surprised that this is an *accumulating* process. Its ultimate rate of change is determined by the input to the inner loop. The delay can be found from the data: $\tau = 6.2$. (A technical detail: before the step test, the queue should not be empty; if it is, then transients from the initial loading of the queue will obscure the steady-state behavior.)

Results from the step test can now be employed to find parameters for the controller in the outer loop, using entries for the AMIGO method for accumulating processes from Table 9-1. The resulting gains are (approximately)

$$k_p = 0.06 \qquad k_i = 0.001 \qquad k_d = 0.2$$

Typical results when using a two-term (PI) controller are shown in Figure 16-3. The setpoint is held constant at $r = 200$, and the intensity of incoming traffic changes twice over the course of the experiment. (The final traffic is about 20 percent higher than the original one.) We see that the external controller is able to maintain the setpoint in the steady state but that the system does not respond well to changes. As

---

1. The data was collected from a simulated system. We will show and discuss the simulation code later in this chapter.

the traffic intensity decreases, the number of server instances is not reduced quickly enough to prevent the queue from running empty for about 50 time steps. Much worse is the behavior as the traffic intensity suddenly increases: by the time enough server instances are online to handle the new load, the queue has overshot the setpoint by a factor of almost 7!

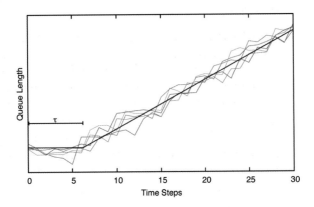

Figure 16-2. Results of step tests performed on the inner loop.

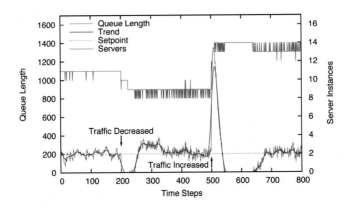

Figure 16-3. Behavior of the outer loop under PI control with k_p = 0.06 and k_i = 0.001.

Derivative Control to the Rescue

The reason for the poor performance is clear: the external controller does not respond adequately to persistent deviations of the queue length from the selected setpoint. However, we cannot increase the integral gain without bringing about violent control oscillations. But there is another term in the proverbial "three-term" or PID controller that responds to *changes* in the tracking error: the derivative term.

We have not used derivative control much (or at all) in this book, given that the derivative term is susceptible to noise (see Chapter 4) and that most of the systems we have considered are noisy. In this example the queue length is also noisy, so why do we believe that derivative control might be applicable now when it wasn't before? The reason is that we have *decoupled* the time scales. Because fast response to high-frequency noise is handled by the inner loop, the external loop can respond more slowly. In particular, we can apply a smoothing filter to remove most of the noise and then use derivative action to bring about a quicker response to changes in the queue length.

Figure 16-4 is typical for the best behavior that can be obtained in this way. The derivative action is filtered using a recursive filter with $\alpha = 0.15$ (see Chapter 10), and the controller gains are adjusted manually from the settings suggested by the AMIGO tuning method. The gains used were

$$k_p = 0.35 \qquad k_i = 0.0025 \qquad k_d = 4.5$$

The performance is much improved over the behavior without derivative action. The interval where the queue is running empty as the traffic intensity decreases has been almost eliminated, and the overshoot in response to the increase in traffic has been reduced by more than half.

Finally, observe that the number of server instances changes more frequently than before. This is a consequence of the derivative term, which tends to enhance high-frequency noise. An excessive amount of control actions is usually undesirable, so we may want to insert a smoothing filter between the external controller and the inner loop. Such a filter will stabilize the setpoint experienced by the inner loop. Figure 16-5 shows what happens after a recursive filter with $\alpha = 0.5$ has been added: the number of server instances fluctuates less, yet the behavior of the queue length is hardly affected at all. Notice that we can apply only a moderate amount of smoothing. As soon as we begin to make α smaller (thereby enhancing the effect of the filter), the queue length begins to drift.

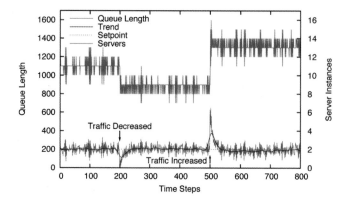

Figure 16-4. Behavior of the outer loop when a derivative term is included in the K_1 controller.

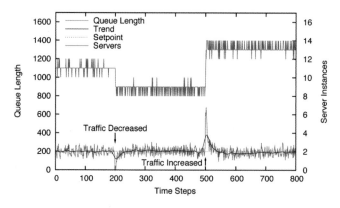

Figure 16-5. Similar to Figure 16-4, but now an additional smoothing filter has been inserted between the external controller and the inner loop.

Controller Alternatives

The nested loop arrangement described here involves two controllers, one each for the outer and the inner loop. For the present solution, we have chosen to use controllers of the PID (or PI) type for both. The resulting performance may be good enough, but it is worth considering alternatives.

The inner loop is suffering from the defect—encountered already in Chapter 15—that you can't have "half a server." If the number of servers required to maintain a particular setpoint is not a whole integer, then the system will end up oscillating rapidly between the two neighboring states. To avoid this artifact, we might want to consider a controller, like the one discussed in Chapter 15, that knows about the integer constraint and switches the number of server instances less frequently. Such a controller will typically exhibit some form of hysteresis in that the number of servers is not changed unless the reasons for a change become sufficiently strong. Such an arrangement will reduce the number of control actions, but it will also decrease the system's tracking accuracy because hysteresis will prevent the system from responding quickly to changes in control input. It will be an engineering decision to balance these requirements.

The controller in the external loop maintains the overall queue length. In the current arrangement, the controller attempts to keep the queue at one specific setpoint *value*, but it may be more appropriate to keep the queue within a particular *range*. This would require a controller that generates little (or no) control action if the tracking error is small but that generates a disproportionately large corrective action whenever the tracking error leaves the desired range. One could try an "error-square" controller (see Chapter 10) or invent a special-purpose controller for this purpose. Whether a special-purpose controller is necessary for a particular application is, again, an engineering decision. One argument generally favoring PID controllers is their simplicity and the relatively small number of adjustable parameters they require.

Simulation Code

The simulation for the queueing system makes use of the Abstract ServerPool base class that was introduced in Chapter 15. (Code discussed previously is not reproduced here.) The main difference between the `QueueingServerPool` used here and the `ServerPool` used in Chapter 15 is that, in the present case, the new load is *added* to the existing queue instead of replacing it.

```
class QueueingServerPool( AbstractServerPool ):
    def work( self, u ):
        load = self.load()      # additions to the queue
        self.queue += load      # new load is added to old load

        completed = AbstractServerPool.work( self, u )

        return load - completed # net change in queue length
```

Recall that the `AbstractServerPool` does not model individual requests. Instead, we specify two functions to place an aggregate "load" onto the queue and one that simulates the "work" done by the servers. This implies that no information regarding individual requests (such as their arrival time or age) is available. Hence the implementation of the `QueueingServerPool` just described reports only the net change in queue length as process output.

The `QueueingServerPool`, when used in a closed-loop arrangement with a PI controller, forms the inner loop of the overall architecture (compare Figure 16-1). As far as the outer loop is concerned, the entire inner loop is just another component: it has a control input and a control output. We therefore model the inner loop as a subclass of the general `Component` abstraction from the simulation framework:

```
class InnerLoop( fb.Component ):
    def __init__( self, kp, ki, loader ):
        k = 1/100.

        self.c = fb.PidController( kp*k, ki*k )
        self.p = QueueingServerPool( 0, consume_queue, loader )

        self.y = 0

    def work( self, u ):
        e = u - self.y       # u is setpoint from outer loop
        e = -e               # inverted dynamics
        v = self.c.work( e )
        self.y = self.p.work( v ) # y is net change
        return self.p.queue

    def monitoring( self ):
        return "%s %d" % ( self.p.monitoring(), self.y )
```

To run the simulations, we need a setpoint that changes over time as well as a load function that models varying intensity of request traffic.

(The load_queue() function must use a global variable to keep track of the current time step.)

```python
def load_queue():
    global global_time
    global_time += 1

    if global_time > 2500:
        return random.gauss( 1200, 5 )

    if global_time > 2200:
        return random.gauss( 800, 5 )

    return random.gauss( 1000, 5 )


def setpoint( t ):
    return 200

    if t < 2000:
        return 100
    elif t < 3000:
        return 125
    else:
        return 25
```

With all these definitions now in place, we can give the code for the simulation used to produce the data shown in Figure 16-4. (Notice the additional filter inserted as "actuator" between the external controller and the inner loop.)

```python
import feedback as fb

fb.DT = 1

global_time = 0 # To communicate with load_queue functions

p = InnerLoop(0.5, 0.25, load_queue) # "plant" for outer loop
c = fb.AdvController( 0.35, 0.0025, 4.5, smooth=0.15 )

fb.closed_loop( setpoint, c, p,
                actuator=fb.RecursiveFilter(0.5) )
```

CHAPTER 17

Case Study: Cooling Fan Speed

A "classic" application of feedback principles is provided by the automatic adjustment of cooling fan speeds in order to maintain some equipment at a desirable temperature—for example, the CPU in a computer or laptop. This system was already introduced in Chapter 5. In contrast to most of the other examples discussed, in this case the governing laws are known at the outset. This shifts the focus of our investigation: rather than trying to obtain a basic, approximate description of the dynamics, we need to find numerical values for the parameters of an existing model.

The Situation

We want to control the speed of cooling fans to maintain a desired temperature of the cooled component. The control output is the temperature, the control input is the fan speed, which is adjustable continuously so that we can treat it as a floating-point number. The heat generated by the CPU depends on its "load," which we will model as changing by fixed steps at random intervals. We will also assume that the ambient temperature may undergo slow, random drifts. Besides these two effects, the system is essentially deterministic.

As mentioned in Chapter 5, the dynamics we wish to control are the *cooling* dynamics—that is, the reduction in temperature as the fan speed is increased. The initial state, where the system is considered to be "off," therefore corresponds to the situation with minimal cooling (with the fan speed reduced to the lowest possible speed that won't damage the CPU). This initial state does *not* correspond to the situation where the CPU is itself switched off, as we are not concerned about the dynamics of the CPU heating up after first being powered on.

The Model

The temperature of a body losing heat to the environment is described by Newton's law of cooling, $\frac{d\Theta}{dt} + \frac{1}{T}\Theta = cu$. Here Θ is the temperature *difference* between the body and its environment:[1]

$$\Theta = \Theta_{body} - \Theta_{ambient}$$

and u describes any heat supplied to the body from outside "sources."

To understand what this differential equation tells us, it is convenient to rewrite it in the following form:

$$\frac{d}{dt}\Theta = -\frac{1}{T}\Theta + cu$$

The change in temperature (per unit of time) $\frac{d\Theta}{dt}$ consists of two contributions, a *loss* of temperature $-\frac{1}{T}\Theta$ and a *gain* of cu. Let's first consider the case where no heat is supplied: $u = 0$. A cup of coffee cooling on the desk is an example. In this situation, the body loses a certain fraction of its temperature every moment. The parameter T is the time scale of the problem: the length of time it takes for the temperature to drop to about one-third of its original value.

The other term on the right-hand side describes any heat supplied to the body. In the CPU example, this is the heat generated by the chip as it is operating. The quantity u is the flow of heat to the body, measured in joules per second (or watts). As the load on the processor changes, so will the amount of heat u generated. The coefficient c describes by how many degrees the body heats up for each joule of energy supplied (basically, c is the heat capacity of the body). You can convince yourself that cu has the dimension of temperature/time.

But where is our fan speed? It is there, hiding inside the quantity T: if the fan runs faster, then the processor will take less time to shed the

---

1. We use the letter Θ for the temperature in order to reserve the letter T for the time constant of the process.

same amount of heat and so T will be smaller. The way the control action enters the equation is a bit sneaky: usually, the control action would be a linear term on the right-hand side, like u. (In fact, the same equation describes a pot that is heated, in which case the control action u is the supplied heat.)

We can now collect all the pieces. If initially the body is at temperature $\Theta(t)$, then its temperature a short time δt later is

$$\Theta(t+\delta t)=\Theta(t)+\delta t\left[-\frac{1}{T}\Theta(t)+cu(t)\right]$$

Here we have made use of the fact that the derivative can be approximated as a finite difference:

$$\frac{d\Theta}{dt}\approx\frac{\Theta(t+\delta t)-\Theta(t)}{\delta t}$$

This approximation is good provided that δt is sufficiently small—in other words, as long as Θ does not change much during an interval of duration δt. For us, this means that δt must be much smaller than T (since T is the time scale over which Θ changes significantly).

As long as T and u are held constant, we can find an explicit solution to the differential equation. Under the stated conditions, the temperature at time t is given by

$$\Theta(t)=\Theta_0 e^{-t/T}+cuT$$

where the constant Θ_0 is determined by the initial temperature. For instance, if the initial temperature is zero ($\Theta(0)=0$) then Θ_0 must be $-cuT$. This describes the situation when the computer is initially switched off and is being turned on at $t=0$. (Remember that Θ is the temperature *above* the ambient one.)

Finally, we need to find numerical values for the various parameters. The power consumption of a current processor is about 75 watts, and I will assume that the power can change in steps of 10 watts as the load changes. The parameters T and c must, in principle, be measured. Here, I estimate $T=120$ seconds. That is to say: *without* the fans running, the processor will drop from 200 degree Celsius to about 70 degree Celsius within 2 minutes; it will cool down faster with the fans running. We can then find c from the final, steady-state temperature that the processor reaches without active cooling. In this limit, $\Theta(\infty)$ = cuT. Assuming that the maximum temperature is 200 degrees Celsius, and plugging in our estimates for u and T, we find $c = \frac{1}{45}$ degrees Celsius per joule.

Tuning and Commissioning

Figure 17-1 shows the kinds of measurements we would perform to determine the values of the parameters.[2] The graph includes curves for several different types of open-loop measurement. One curve shows the temperature development for the CPU *without* any of the fans running. The temperature tops out somewhere near 200 degrees, at which point the chip has overheated and serves only as a space heater but is useless for any other purpose!

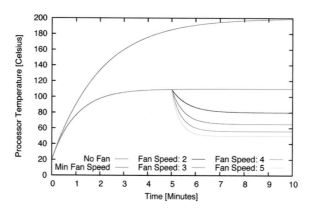

Figure 17-1. Time development of the CPU temperature in an open-loop arrangement. The fan speed is suddenly increased from its minimal setting to various higher speeds at t = 5 minutes, resulting in greater cooling action and a decrease in temperature (step test).

Let's suppose the maximal operating temperature that won't damage the chip is 100 degrees. Maintaining this temperature requires the fans to be running, but at minimum speed. At t = 5 minutes, we increase the fan speed: that's basically the kind of "step test" discussed in Chap-

---

2. The data was collected from a simulated system. We will show and discuss the simulation code later in this chapter.

ter 8, and from the graph we can read off the decrease in temperature due to this control action. (Note that the decrease in temperature is not a linear function of the fan speed.)

The figure shows that the delay τ is negligible and that the time constant T is approximately 1 minute (60 seconds). At fan speed 4, the temperature reduction is about 40 degrees and so the static gain factor $\Delta u/\Delta y$ is approximately 4/40 = 0.1. Quite satisfactory closed-loop performance is found with $k_p = 2.0$ and $k_i = 0.5$.

Closed-Loop Performance

In Figure 17-2 we see how this system behaves in production. The setpoint is initially set to 50 degrees and is later reduced to 45 degrees. The load on the processor keeps changing, and with it the amount of heat generated. But whenever the amount of heat increases, the fan speed increases with it to keep the CPU temperature at the setpoint. The temperature overshoots a little whenever the load level changes, but it quickly reaches the desired setpoint again.

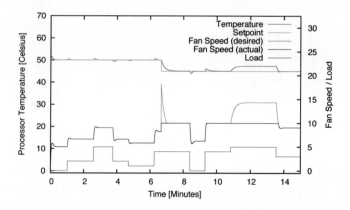

Figure 17-2. Time development of the CPU temperature under closed-loop control. The fan is not capable of delivering the required cooling actions if the CPU is working under maximum load (actuator saturation, t = 11, ..., 14).

One practical problem exhibited by the system is the occurrence of *actuator saturation*. The fans are not capable of keeping the CPU at 45 degrees at the highest load level, with the consequence that the desired fan speed (as calculated by the controller) is higher than the actual speed that the fan can achieve. It is therefore important to use a "clamping" controller, which stops adding to the integral term once the actuator is maxed out (see Chapter 10).

Simulation Code

The dynamics of the system in the present case study are described by a differential equation; simulating it therefore means solving (or "integrating") this differential equation. In the code that follows, this is done in the simplest possible manner. We take the equation in the form

$$\Theta(t+\delta t)=\Theta(t)+\delta t\left[-\frac{1}{T}\Theta(t)+cu(t)\right]$$

and translate it directly into code. This equation amounts to an updating algorithm for $\Theta(t)$: it uses the temperature at time t to calculate the temperature at a later time $t+\delta t$. The quantity T, which describes how quickly the CPU cools down, is a combination of the natural heat loss and the heat loss due to active cooling from the fan. The term $u(t)$, which describes the heat generated by the CPU, is a combination of the chip's idle power and the additional heat if the CPU is running under increased load. The load itself changes at random times by a fixed amount.

```
import random
import feedback as fb

class CpuWithCooler( fb.Component ):
    def __init__( self, jumps=False, drift=False ):
        self.ambient = 20              # temperature: degree C
        self.temp    = self.ambient    # initial temperature

        self.wattage = 75              # CPU heat output: J/sec
        self.specific_heat = 1.0/50.0  # specific heat: degree/J

        self.loss_factor = 1.0/120.0   # per second

        self.load_wattage_factor = 10  # addtl watts due to load
        self.load_change_seconds = 50  # avg sec between changes
        self.current_load = 0

        self.ambient_drift = 1.0/3600  # degrees per second

        self.jumps = jumps             # jumps in CPU load?
        self.drift = drift             # drift in ambient temp?
```

```
def work( self, u ):
    u = max( 0, min( u, 10 ) )         # actuator saturation

    self._ambient_drift()               # drift in ambient temp
    self._load_changes()                # load changes, if any

    diff = self.temp - self.ambient # temperature diff
    loss = self.loss_factor*(1 + u) # natural heatloss+fan

    flow = self.wattage + self.current_load # CPU heat flow

    self.temp += fb.DT*(self.specific_heat*flow - loss*diff)
    return self.temp


def _load_changes( self ):
    if self.jumps == False: return

    s = self.load_change_seconds
    if random.randint( 0, 2*s/fb.DT ) == 0:
        r = random.randint( 0, 5 )
        self.current_load = self.load_wattage_factor*r

def _ambient_drift( self ):
    if self.drift == False: return

    d = self.ambient_drift
    self.ambient += fb.DT*random.gauss( 0, d )
    self.ambient = max( 0, min( self.ambient, 40 ) )


def monitoring( self ):
    return "%f" % ( self.current_load, )
```

The units we use in this case study to measure wall-clock time are
seconds, and all other units are compatible with this scale (for example,
watts equal joules per second). As we saw previously, the typical time
scale on which the system's temperature changes significantly is about
1 minute. We therefore need to take steps that are at least one order of
magnitude smaller than that—about a second or less. In the end, I used
steps of one-hundredth of a second (fb.DT = 0.01) for extra accuracy,
although one-tenth of a second (fb.DT = 0.1) would have been sufficient.

One important aspect that this example aims to demonstrate is the
effect of actuator saturation: there is a maximum cooling effect the fan
can achieve, because at some point it simply can't run any faster. The
output of the fan is even more constrained in the opposite direction,

because the fan output can never become *negative* (corresponding to a heating effect).

To model this behavior, a Limiter element has been included in the position of an *actuator* between the controller and the plant. This element constrains its output to the range that was specified when the element was first created, and any input that exceeds this range is trimmed to the most extreme value still within the permitted range.

```python
class Limiter( fb.Component ):
    def __init__( self, lo, hi ):
        self.lo = lo
        self.hi = hi

    def work( self, x ):
        return max( self.lo, min( x, self.hi ) )
```

Given the effect of this Limiter, it is important to use a *clamping* controller. So instead of the simple PidController, we use an AdvController instance, where the clamping range is set equal to the range permitted by the actuator.

```python
fb.DT = 0.01

def setpoint(t):
    if t < 40000: return 50
    else: return 45

p = CpuWithCooler( True, True ); p.temp = 50 # Initial temp
c = fb.AdvController( 2, 0.5, 0, clamp=(0,10) )

fb.closed_loop( setpoint, c, p, 100000, inverted=True,
                actuator=fb.Limiter( 0, 10 ) )
```

CHAPTER 18

Case Study: Controlling Memory Consumption in a Game Engine

This final case study allows us to see how a nonproportional controller can be used in a feedback loop to good effect.

The Situation

Imagine a game engine managing the movements of various game objects (widgets, sprites, space ships, whatever). During the course of a game, the number of game objects changes. Our job is to make sure that the memory consumption of the game engine stays within acceptable limits even as the number of game objects grows large.

The way to control memory consumption in the game engine is via the graphics resolution: high resolution translates into high memory consumption. To make things concrete, let's say that we can choose from five different resolution levels and that each subsequent level requires twice as much memory per game object. (For instance, the levels may set aside 100, 200, 400, 800, and 1,600 memory units per game object.) The problem of choosing the correct resolution therefore appears quite simple: divide the maximum amount of memory available by the number of game objects to find the greatest number of memory units available to each object, and then choose the greatest resolution that respects this limit. We don't need feedback for that.

Remember that feedback is a method to achieve robust control in the face of uncertainty. Feedforward works great as long as the system is totally deterministic and all relevant information is available. So in this case, if the amount of memory required can be predicted accurately from the number of game objects, then there is no need for feedback control. But memory usage may be more complicated than that. In a contemporary, managed programming environment that includes a garbage collector, the amount of memory allocated may change in its own idiosyncratic ways. There may be other sources of uncertainty, such as leaks, overhead, sharing, and caching. Thus the actual amount of memory used can change quite unpredictably. And all of a sudden, feedback seems like a good idea.

Problem Analysis

The control task in this case differs from those we have discussed before. The two most important differences are the following:

- We do not care about tracking a setpoint accurately. Instead, we want to prevent the process output from leaving a specified interval.

- The number of input values is extremely limited, and so proportional control is out of the question: the controller must choose one of the available levels. (In fact, the levels may not even be numerical—they might be *categorical* labels, such as "high," "medium," and "low.")

Taken together, these two points suggest a controller that exhibits a central dead zone but produces a steplike response once the output variable leaves the allowed interval (see Figure 18-1). The dead zone should be relatively wide so that changes in graphics resolution are rare.

The only control actions available to us are to increase or decrease the resolution level. At this point, we must decide how to encode those. Rather than dealing with actual values of the memory consumption per game object (which may change from platform to platform and may not even be known), we will simply indicate the resolution by an integer between 0 and 4, with the lowest resolution (and the lowest memory consumption) corresponding to the index 0. The numerical values carry no significance (they are essentially categorical labels), but it is important that they exhibit an *ordering* relation so that there is a well-defined directionality to the input/output relation. With these conventions, increasing the control input (the level) leads to an increase in process output (the memory consumption).

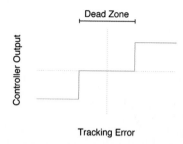

Figure 18-1. The input/output relationship for a dead-zone controller. The controller output is zero unless the tracking error is greater than the width of the dead zone.

We can now interpret the controller output as increments of the resolution level: a positive output will increase the resolution level by 1; a negative output will decrease the resolution level by the same amount. To keep track of the current resolution, we introduce an aggregator (or integrator) into the control loop. This element maintains the current resolution as its internal state and increases or decreases it, depending on its input. In addition, the aggregator ensures that the resolution level is valid: it will neither decrement past 0 nor increment past 4. (This is an example of "actuator saturation"; see Chapter 10.)

In any case, we assume that the game engine responds immediately to any request for a change in resolution, so there are no nontrivial dynamics that need to be taken into account.

Architecture Alternatives

There are two different ways to turn these general considerations into a concrete loop architecture:

- The first alternative forces us to think about feedback architectures in a different way, because it forgoes many of the usual elements of a control loop, including such essentials as the setpoint and the tracking error.

- The second loop arrangement stays much closer to the familiar setup, but it requires some mathematical sleight of hand.

A Nontraditional Loop Arrangement

The first way to think about the loop architecture is shown in Figure 18-2 (top). The "setpoint" now consists of the allowed range, which is specified by a pair of numbers (the upper and lower limits of the range). Since the reference signal is no longer a single value, it does not make sense to form the difference between the reference and the process output. Instead, both the allowed range and the process output are fed to the controller. The controller determines whether or not the process output falls within the allowed range, and it produces an output value based on that condition. Because no difference between output and signal is ever calculated, there is no need to multiply the output by –1 on the return path.

The controller has no adjustable parameters, so there is no need for any tuning. The operation of the loop is completely determined by the range supplied through the reference signal.

A Traditional Loop with Logarithms

The other possible loop architecture (Figure 18-2, bottom) retains the concept of a scalar setpoint value and the tracking error as the difference between the setpoint and the process output. Yet because the memory allocated per game object changes by a constant *factor* from one level to the next, it will be necessary to base the control actions on the *logarithm* of the respective signals. To give an example: if we switch from the lowest resolution to the next one, then the memory consumption per game object changes from 100 to 200 units—a difference of 100 units. The next resolution level assigns 400 units per game object, which is a difference of 200. It therefore seems as if not all steps are equal: the higher the resolution, the further apart the levels are spaced. However, if we use logarithms of the memory consumption then the distance between consecutive levels is always the same:

$$\log 200 - \log 100 = \log \frac{200}{100} = \log 2$$

$$\log 400 - \log 200 = \log \frac{400}{200} = \log 2$$

$$\vdots$$

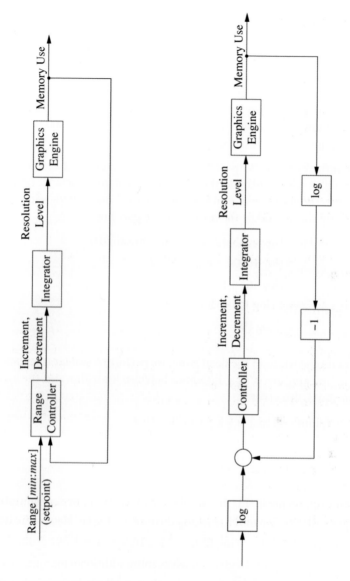

Figure 18-2. Two possible loop architectures for this case study. In the less traditional arrangement (top), the reference signal consists of a pair of numbers that specify the allowed range. No tracking error is calculated; the controller determines whether the process output falls within the allowed range. In the more classical loop (bottom), the reference is a scalar value that is the logarithm of the midpoint in the allowed range.

For this reason, we will use the logarithm of the memory consumption and the setpoint when we calculate the tracking error.

Using this loop architecture, the permissible range is specified as fixed controller parameters when the controller is first set up. Our requirements are that the controller should produce a nonzero output if memory consumption exceeds (say) 10,000 units. We will also introduce a lower threshold: if memory falls below (say) 1,000 units, the resolution should be increased.

We now choose as "setpoint" the arithmetic middle between the two threshold values and make the dead zone equal to the distance between the thresholds. Keep in mind that we are working with logarithms, so the setpoint should be the mean of the logarithms:

$$\begin{aligned}
\text{setpoint} &= (\text{upper threshold} - \text{min threshold})/2 \\
&= (\log 10000 + \log 1000)/2 \\
&= (\log 10^4 + \log 10^3)/2 \\
&= (4\log 10 + 3\log 10)/2 \\
&= \tfrac{7}{2}\log 10
\end{aligned}$$

To find the width of the dead zone, we perform a similar calculation. The "tracking error" is a difference between logarithms. At the lower threshold, this error is

$$\begin{aligned}
\text{setpoint} - \text{actual} &= \tfrac{7}{2}\log 10 - \log 1000 \\
&= \tfrac{7}{2}\log 10 - 3\log 10 \\
&= \tfrac{1}{2}\log 10
\end{aligned}$$

At the upper threshold (that is, when the actual memory consumption equals 10,000), we find a tracking error of $-\tfrac{1}{2}\log 10$. Hence, the dead zone extends from $-\tfrac{1}{2}\log 10$ to $+\tfrac{1}{2}\log 10$, symmetrically around the setpoint at $\tfrac{7}{2}\log 10$. To give ourselves some additional margin, we can choose tighter bounds, of course, such as $\pm\tfrac{1}{2}\log 8$, for example.

Results

Typical results are shown in Figure 18-3.[1] Memory consumption increases with the number of game objects. As consumption exceeds the critical threshold, the resolution level is decremented with the consequence that the memory required drops sharply (in fact, it is cut in half). This is especially obvious in the second half of the simulation: the number of game objects increases steadily; however, whenever the total memory consumption approaches the upper threshold, the resolution level is reduced and with it the amount of memory consumed. Similarly, if the number of game objects becomes small, then the resolution level is increased accordingly.

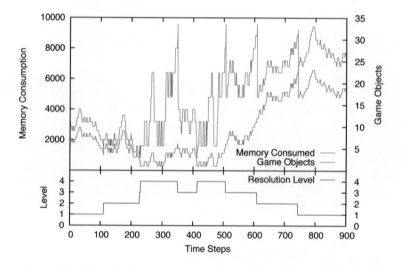

Figure 18-3. Typical closed-loop performance. In general, the amount of memory consumed grows with the number of game objects. If the memory consumption approaches the upper threshold, then the resolution level is reduced, cutting the amount of occupied memory in half.

1. The data was collected from a simulated system. We will show and discuss the simulation code later in this chapter.

Simulation Code

In the spirit of the simulation framework, the following code uses the loop architecture based on a scalar setpoint (Figure 18-2, bottom). It is not difficult, however, to adapt the code to follow the nonstandard architecture in Figure 18-2 (top).

The simulation code describes a system with deterministic memory usage, but it is easy to add further sources of randomness. The controlled system itself is described by the class GameEngine. At each time step, there is a possibility that the number of game objects will change by 1 (in either direction). The total number of game objects must fall between 1 and 50.

```python
import math
import random
import feedback as fb

class GameEngine( fb.Component ):
    def __init__( self ):
        self.n = 0      # Number of game objects
        self.t = 0      # Steps since last change

        # for each level: memory per game obj
        self.resolutions = [ 100, 200, 400, 800, 1600 ]

    def work( self, u ):
        self.t += 1

        # 1 change every 10 steps on avg
        if self.t > random.expovariate( 0.1 ):
            self.t = 0
            self.n += random.choice( [-1,1] )
            self.n = max( 1, min( self.n, 50 ) ) # 1 <= n <= 50

        crr = self.resolutions[u] # current resolution
        return crr*self.n            # current memory consumption

    def monitoring( self ):
        return "%d" % (self.n,)
```

In addition to the game engine, we need several other components to complete the control loop: the controller itself, the aggregator or actuator (called ConstrainingIntegrator in the code because it also constrains its output to the range of legal resolution levels), and finally an element to calculate the logarithm of the output signal.

```python
class DeadzoneController( fb.Component ):
    def __init__( self, deadzone ):
```

```
        self.deadzone = deadzone

    def work( self, u ):
        if abs( u ) < self.deadzone:
            return 0

        if u < 0: return -1
        else:     return 1


class ConstrainingIntegrator( fb.Component ):
    def __init__( self ):
        self.state = 0

    def work( self, u ):
        self.state += u
        self.state = max(0, min( self.state, 4 )) # Allow 0..4
        return self.state


class Logarithm( fb.Component ):
    def work( self, u ):
        if u <= 0: return 0
        return math.log(u)
```

With all these components in place, we need only define a function to provide the setpoint; we can then complete the closed-loop operation in a single call.

```
fb.DT = 1

def setpoint(t):
    return 3.5*math.log( 10.0 )

c = DeadzoneController( 0.5*math.log(8.0) ) # width of deadzone
p = GameEngine()

fb.closed_loop( setpoint,c,p,actuator=ConstrainingIntegrator(),
                returnfilter=Logarithm() )
```

CHAPTER 19

Case Study Wrap-Up

In wrapping up this collection of case studies, I'd like to point out a few recurrent themes.

Simple Controllers, Simple Loops

The elements of a feedback system are not complicated: the basic loop structure and a simple controller are all that is required. In fact, *all* the case studies were completed with nothing more complicated than the generic feedback loop as implemented in the closed_loop() convenience function!

Most controllers, also, were of the generic PID type, although particular situations sometimes called for specially designed controllers. But even those controllers were very simple and did nothing more than calculate an output based on the input while maintaining only minimal internal state.

The natural temptation to build more complicated controllers should probably be resisted in most cases. The feedback principle is not about clever (and complex) algorithms; rather it works with simple components put together in a straightforward fashion. What makes feedback work is that corrective actions are calculated and applied *constantly*. Because of the iterative nature implied by the feedback scheme, the components and calculations themselves can (and should) be simple.

Measuring and Tuning

Given all the details and specific methods to "measure the transfer function" presented in Chapter 8, it is easy to forget that we are really after only a few pieces of basic information:

- What is the *directionality* of the input/output relationship? Does increasing the control input result in an increase or a decrease of the process output?
- What is the typical *time scale T* of the process? How long does it take the output to settle to a new steady state again after a sudden change in control input?
- Is there a significant *delay τ* before a change in input becomes observable in the output?
- What is the *static gain factor Δu/Δy*? How much do we need to change the input to bring about a permanent change in output?

All the methods presented earlier are just ways to obtain those basic pieces of knowledge about the process in question.

The same can be said for controller tuning. For PID controllers, the controller gains consist of the static gain factor $\Delta u/\Delta y$, which is modified by a factor that takes the dynamic response of the process into account (increasing the gain for sluggish processes, decreasing the gain when there is noticeable delay). Given only those bits of process knowledge, we can find controller gains that will result in a workable closed-loop operation and that can be improved through trial and error in a manual process. The formulas and methods discussed in Chapter 9 are primarily shortcuts for this process.

Staying in Control

There are a variety of signals flowing around a control loop: the setpoint, the process output, the tracking error, the controller output. Add some additional elements, like actuators or filters, and we are talking about half a dozen individual signals for a basic loop alone! In a nested arrangement, the number of signals multiplies.

It is surprisingly easy to get confused about which signal goes where, and with which sign (plus or minus). If a newly commissioned loop does not seem to work *at all*, it often helps to trace all signals around the loop and to confirm that the components were indeed wired together correctly. To ensure that the signs are correct it helps to ask, for each component: if the input goes up, should the output go up or down—and what does it actually do?

Dealing with Noise

Most of the case studies showed systems that included a stochastic aspect. In fact, randomness and the uncertainty that it brings about will often be what makes feedback control an attractive proposition in the first place.

In all the case studies, I have taken a "naive" approach to noise. In essence we assumed that we could ignore the randomness, and concentrate only on the deterministic part of the system, on the premise that the noise would "average itself out" over time. In a similar spirit, we have usually shown only a single simulation run for each system with the understanding that the observed behavior is typical and a representative sample from all possible runs. Performing several simulation runs quickly gives one a sense for the magnitude of the variations that can be expected. If greater accuracy is required (in simulations or a production installation), then one can calculate the desired quantity as the average over several experimental runs.

It is a natural impulse to use filters as elements in a control loop to obtain smoother signals. Although that often makes sense, it is not always necessary. Keep in mind that filters slow signals down; therefore what we gain in smoothness via filtering may be lost again because the system now requires greater controller gains. Filters should be used only when they have been shown to be necessary. One should not automatically reach for a filter just because a signal is noisy.

Finally, all the case studies assumed that the noise involved was relatively harmless: possibly of large amplitude but always of finite variance. This is not necessarily so: systems exhibiting noise with a power-law spectrum do exist and always require special treatment.

PART IV

Theory

The material that follows is not strictly required reading—in particular, when you are mostly interested in practical applications of feedback techniques to computer systems.

At the same time, it is not recommended to skip these chapters entirely. Many operations and concepts will make more sense once their theoretical underpinnings are understood. And all this material is simply required if you ever want to dig further into feedback and control theory.

But feel free to skim along, and to come back to dive deeper as the need arises.

CHAPTER 20

The Transfer Function

As Chapter 3 demonstrated, understanding a system's dynamic behavior is important for building a stable and well-performing feedback
loop. In this chapter, we will first describe how to capture information
on a system's dynamic behavior; we then show how to repackage this
information in a way that is particularly convenient for our purposes.
The tool that we will use is the *transfer function*.

Differential Equations

The usual way to describe the time evolution of a system is through
differential equations. A differential equation is an expression involving the derivative of a quantity, often together with the quantity itself.
Here are some examples of differential equations:

$$\frac{d}{dt}y(t) + \frac{1}{T}y(t) = u(t) \qquad \text{Newton's law of cooling}$$

$$\frac{d^2}{dt^2}y(t) + 2\zeta\omega\frac{d}{dt}y(t) + \omega^2 y(t) = f(t) \quad \text{Damped harmonic oscillator}$$

$$\frac{d}{dt}y(t) - a\big(y(t)\big)^2 = 0 \qquad \text{Riccati equation}$$

Because the derivative is the rate of change of the quantity, differential
equations are the natural way to describe how a system changes over
time: they describe the system's *dynamics*. "Solving" a differential
equation means finding a curve $y(t)$ that, for all times t, fulfills the
differential equation. Several analytical and numerical methods exist
to find the solution to a given differential equation.

Laplace Transforms

Differential equations provide an especially compact way of describing the dynamics of a system: all possible trajectories, for all times t, can be obtained from the differential equation alone.[1] We now repackage this information in a way that makes it easier to manipulate. Rather than considering the quantities of interest as functions of time, we transform them to the *frequency domain* or *frequency space*.

To begin with, consider an arbitrary well-behaved function $f(t)$. We can calculate its *Laplace transform $F(s)$* as

$$F(s) = \int_0^\infty f(t)\, e^{-st}\, dt \qquad s \in \mathbb{C}$$

where s is an arbitrary *complex* number. Note that $f(t)$ is a function of t whereas $F(s)$ is a function of s. Since t has dimensions of time and since the exponent of the exponential must be dimensionless, it follows that s must have dimension of 1/time; for this reason, we will refer to it as *frequency*. Functions of t are said to be in the *time domain*, whereas their Laplace transforms are in the frequency domain.

As it turns out, the quantity $F(s)$ contains exactly the same information as $f(t)$. We can regain the original function by means of an inverse transformation:

$$f(t) = \frac{1}{2\pi i} \int_{\sigma - i\infty}^{\sigma + i\infty} e^{st} F(s)\, ds \qquad \sigma \in \mathbb{C}$$

In practice, one rarely finds the Laplace transform of some function (or its inverse) by explicitly evaluating the respective integrals. Transform pairs for most commonly used functions can be looked up in appropriate tables. One can also establish several generally applicable rules to extend the tabulated results to more general cases. (Some transform pairs and rules are shown in Table 20-1. All of these relations are easily verified by performing the transform integral. Every book

---

1. To pick out one specific trajectory from all possible ones, one also needs to specify its value for $t = 0$, the *initial condition*.

on control theory will contain such tables, often with many more entries, but Table 20-1 will suffice for our purposes.)

Table 20-1. Some commonly used Laplace transform pairs.

Time Domain $f(t)$	Frequency Domain $F(s)$
1 (unit step at $t = 0$)	$1/s$
t	$1/s^2$
t^n	$\dfrac{n!}{s^{n+1}}$
$\dfrac{1}{T}e^{-t/T}$	$\dfrac{1}{1+sT}$
$\dfrac{1}{(n-1)!T}\left(\dfrac{t}{T}\right)^{n-1}e^{-t/T}$	$\dfrac{1}{(1+sT)^n}$
$\cos \omega t$	$\dfrac{s}{s^2+\omega^2}$
$\sin \omega t$	$\dfrac{\omega}{s^2+\omega^2}$
$\dfrac{\omega}{\sqrt{1-\zeta^2}}e^{-\zeta\omega t}\sin\left(\sqrt{1-\zeta^2}\,\omega t\right)$	$\dfrac{\omega^2}{s^2+2\zeta\omega s+\omega^2}$
$f(t)$	$F(s)$
$af(t)+bg(t)$	$aF(s)+bG(s)$
$e^{at}f(t)$	$F(s-a)$
$f(t-T)$	$e^{-Ts}F(s)$ $\qquad T \geq 0$
$f(t/a)$	$aF(as)$
$\dfrac{d}{dt}f(t)$	$sF(s)-f(t=0)$
$\displaystyle\int_0^t f(\tau)\,d\tau$	$\dfrac{1}{s}F(s)$

Properties of the Laplace Transform

Given the definition, it is easy to show that the Laplace transform is *linear*: if $h(t) = a\,f(t) + b\,g(t)$, then its Laplace transform is $H(s) = a\,F(s) + b\,G(s)$ (where a and b are constant scalars). This property allows us to build up the Laplace transform of a more complicated function (such as a polynomial) from the Laplace transform of its components.

However, the essential property of the Laplace transform—and the real reason we care about it to begin with—is the effect it has on *derivatives* and *integrals*.

For example, suppose we want to take the Laplace transform of $\frac{d}{dt}f(t)$, which is the derivative of $f(t)$. From the definition of the Laplace transform, we have

$$\int_0^\infty \tfrac{d}{dt} f(t) \, e^{-st} \, dt = f(t) \, e^{-st} \Big|_0^\infty - \int_0^\infty f(t)(-s)e^{-st} \, dt$$

$$= sF(s) - f(t = 0)$$

where the integration is performed using integration by parts. In other words, taking the Laplace transform of the derivative of a function amounts to multiplying the transform of the function itself by s. (We pick up an additional term for the value of the function at $t = 0$. In what follows, we will always assume that the system is initially "at rest," so that $f(t = 0) = 0$ and thus we can ignore this term from now on.)

For higher derivatives, we obtain an additional factor of s for each order of the derivative. For integrals, however, we pick up a factor of $1/s$—this should not be surprising, since taking the derivative is the inverse operation to performing an integration.

One other operation deserves mention because it occurs frequently in practical applications: if $F(s)$ is the Laplace transform of $f(t)$, then the Laplace transform of $f(t - T)$ is $e^{-sT}F(s)$. In other words, *shifting* the function in the time domain introduces an exponential factor in the frequency domain. Such shifts are common in control problems, since they describe *delays*. (If a system replicates its input $u(t)$ to its output $y(t)$ but introduces a delay of duration T, then the output is a shifted version of the input: $y(t) = u(t - T)$.)

We can summarize our discussion as follows:

- Taking the derivative in the time domain amounts to a multiplication by s in the frequency domain.

- Taking the integral in the time domain amounts to a multiplication by $1/s$ in the frequency domain.

- Shifting the function in the time domain by T to the right introduces a factor e^{-sT} in the frequency domain.

Using the Laplace Transform to Solve Differential Equations

As we have seen, taking the Laplace transform of a derivative replaces each derivative (in the time domain) by a factor of s (in the frequency domain). We can use this property of the Laplace transform to *turn differential equations into algebraic equations*, provided that the differential equation is *linear* and has *constant coefficients*. (A differential equation is linear if the unknown function and its derivatives enter only as linear factors. Of the three differential equations introduced earlier, the first two are linear but the Riccati equation is not—because it contains the term $a(y(t))^2$, which includes the unknown function $y(t)$ raised to the second power.)

A Worked Example

As an example, let's consider the following simple linear differential equation:

$$\frac{dy(t)}{dt} = -\frac{1}{T}y(t) + u(t)$$

This equation describes various decay processes. If we initially ignore the term $u(t)$, then the change in $y(t)$ is proportional to the current magnitude of $y(t)$. Moreover, as long as the constant T is positive, the change is negative. This means that y decays at a constant rate given by $1/T$. A radioactive sample undergoing nuclear decay behaves this way; during each time interval, a specific fraction of the atoms in the sample decay. The differential equation also describes a heated body that is cooling down, because the body loses a fraction of its heat per unit of time.

The term $u(t)$ stands for any type of additional change in $y(t)$ that is independent of the system's internal dynamics. For instance, $u(t)$ might describe heat that is supplied to the body by an external source. What's important is that $u(t)$ is independent from the mechanisms that govern the changes in the system itself: it describes an *external* influence.

It is customary to bring all terms depending directly on the unknown quantity $y(t)$ on the lefthand side, thus:

$$\frac{dy(t)}{dt} + \frac{1}{T}y(t) = u(t)$$

To "solve" this differential equation, we need to find an explicit expression for $u(t)$ that holds for all times t. We will now do so by using the Laplace transform. First, we transform all functions that depend explicitly on time t into functions of frequency s. The differential equation now becomes

$$sY(s) + \frac{1}{T}Y(s) = U(s)$$

Observe that this equation is simply an *algebraic* equation—all derivatives have been replaced by factors of s. Hence we can now factor out the term $Y(s)$ on the left-hand side:

$$\left(s + \frac{1}{T}\right)Y(s) = U(s)$$

and solve for $Y(s)$:

$$Y(s) = \frac{1}{s + (1/T)}U(s) = \frac{T}{1 + sT}U(s)$$

We have now "solved" the differential equation, because we have completed our program to obtain an explicit expression for the unknown quantity. Furthermore, this expression is valid for all possible external influences! The problem is that we have only obtained an expression in the frequency domain. To find the behavior in the time domain, we must transform the expression for $Y(s)$ back; this can be done once the shape of the external influence has been fixed.

The Transfer Function

The solution to the differential equation that we found in the previous section had the following structure:

$$Y(s) = H(s)U(s)$$

Here the term $H(s)$ does *not* depend on the external "forcing" function $U(s)$ in any way: it is completely determined by the differential equation alone. In this way, it encapsulates all the internal dynamics of the system. All the information that is contained in the differential equation about the behavior of the system is now contained in $H(s)$. But because $H(s)$ is simply a function of s, $H(s)$ is easier to work with than using the differential equation directly. The function $H(s)$ is known as the *transfer function* of the system.

Using the Laplace transform, any linear differential equation with constant coefficients can be repackaged into a transfer function. The transfer function can be used to calculate how the system will respond to an arbitrary external influence: just multiply $H(s)$ by the Laplace transform of the influence, and then transform the result back into the time domain. Because the transfer function does not depend on the external forcing function, we can find the response of the system to *any* external influence in this way.

Moreover, it is often not even necessary to perform the back-transformation to the time domain. Much information about the dynamic behavior of the system can be obtained merely from the *structure* of the transfer function alone. In particular, the locations where the transfer function becomes infinite or zero (its *poles* and *zeros*) let us predict how the system will respond to an external disturbance. This will be the topic of Chapter 23 and Chapter 24.

Worked Example: Step Response

Let's work out the step response of the system just described. The differential equation $\frac{d}{dt}y(t) + \frac{1}{T}y(t) = 0$ describes its dynamics, and its transfer function $H(s)$ is

$$H(s) = \frac{1}{1 + sT}$$

which has been normalized, so that its steady-state (or zero-frequency, $s = 0$) gain is unity.

As input to this system we choose a step function $u(t) = 1$ for $t > 0$; according to Table 20-1 this function the Laplace transform

$$U(s) = \frac{1}{s}$$

In frequency space, applying an input to a system amounts to multiplying the system's transfer function by the Laplace transform of the input

$$Y(s) = H(s)U(s) = \frac{1}{1+sT} \cdot \frac{1}{s}$$

$$= \frac{1}{s} - \frac{T}{1+sT}$$

In the second step here we have split the expression into partial fractions. (Just add the two terms together to convince yourself that this is indeed correct.)

The quantity $Y(s)$ is the step response in the frequency domain. In order to find the form of the step response in the time domain, we need to transform this expression back. Using Table 20-1 again, we find that

$$y(t) = 1 - T\frac{1}{T}e^{-t/T} = 1 - e^{-t/T}$$

We can now see why this process is commonly known as "simple lag": the output basically follows the step input but is lagging behind. The response is also simple, without oscillation or other notable behavior.

The simple lag is an extremely important process for practical applications, since it provides the simplest description of any "sluggish" process that basically follows its input. (The process models used in Chapter 8 and Chapter 9 were for the most part based on this type of behavior.)

Worked Example: Ramp Input

In the previous section we found the behavior of a system when subjected to a steplike input. To find the response to a different input, we need only multiply the transfer function by the appropriate input function. For example, if we want to know how the system behaves when we apply a "ramp input" $u(t) = t$, then we must first find the Laplace transform of the ramp; according to Table 20-1, this is $U(s) = 1/s^2$. We now multiply this input by the transfer function:

$$Y(s) = H(s)U(s) = \frac{1}{1+sT} \cdot \frac{1}{s^2}$$

$$= \frac{1}{s^2} - \frac{T}{s} + \frac{T^2}{1+sT}$$

This expression can be transformed back into the time domain, term by term. The final result is

$$y(t) = T\left(\frac{t}{T} - \left(1 - e^{-t/T}\right)\right)$$

This example demonstrates how the transfer function makes it (relatively) easy to find the response to an arbitrary input. Just multiply by the desired forcing function in the frequency domain, and transform back into the time domain.

The Harmonic Oscillator

Another example of considerable practical importance is the harmonic, damped oscillator driven by an external force $f(t)$. It has the following differential equation:

$$\frac{d^2 y(t)}{dt^2} + 2\zeta\omega \frac{dy(t)}{dt} + \omega^2 y(t) = \omega^2 f(t)$$

If we take the Laplace transform of all functions that depend on t while accounting for the effect the Laplace transform has on derivatives, then the preceding differential equation becomes

$$s^2 Y(s) + 2\zeta\omega s Y(s) + \omega^2 Y(s) = \omega^2 F(s)$$

Solving for $Y(s)$ yields

$$Y(s) = \left[\frac{\omega^2}{s^2 + 2\zeta\omega s + \omega^2}\right] F(s)$$

and so the transfer function for the harmonic oscillator is

$$H(s) = \frac{\omega^2}{s^2 + 2\zeta\omega s + \omega^2}$$

This result is of tremendous practical importance because so many mechanical and electrical systems have a tendency to oscillate. The harmonic oscillator serves as a description of (or at least an approximation to) all such systems.

What If the Differential Equation Is Not Known?

The methods developed in this chapter all assume that the differential equation describing the system dynamics is known explicitly. For many mechanical or electrical systems, this is true because the governing laws are well known. For automation processes, however, this is often not the case. Not only may the actual dynamic behavior be entirely unknown, but it may not even be clear what "laws" might describe it—what would take the equivalent place of Newton's laws for, say, a web cache?

Despite these objections, the techniques developed in this chapter (and in those that follow) are still relevant for two reasons. First, even if the dynamics of a system are not known a priori, we can still *measure* its behavior and build a transfer function based on the experimental observations (rather than deriving the transfer function from a differential equation; this was the topic of Chapter 8). Furthermore, the concepts and arguments regarding the dynamic behavior of a system are the same, regardless of whether a differential equation is known explicitly or not.

CHAPTER 21

Block-Diagram Algebra and the Feedback Equation

In Chapter 20 we saw that the dynamic behavior of a system is given as the solution to a differential equation. We also saw how the Laplace transform could be used to repackage all the dynamic information contained in a linear, time-invariant differential equation into a simple function (the *transfer function*). In this chapter, we will show how the dynamic behavior of a *combination* of systems can be found from the transfer functions of the individual elements.

Composite Systems

In Chapter 20, we saw that, in the frequency domain, a system's dynamic response $y(s)$ to an external input $u(s)$ is given by the product of the system's transfer function $H(s)$ and the input[1]

$$y(s) = H(s)u(s)$$

We can express this equation as a *block diagram*, where the system (described by its transfer function H) transforms the input u to the output y:

---

1. When working entirely in the frequency domain (in which case there is no need to have separate designations for quantities in the time domain), it is customary to use lowercase letters for input and output signals and to reserve uppercase letters for elements.

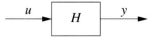

Obviously, we can combine several such systems in series, with the output of one serving as input to the next:

The output of this composite system is the product of its components:

$$y(s) = H(s)\, G(s)\, u(s)$$

This follows simply because the output of the first element is $x(s) = G(s)\, u(s)$ and because the output of the second component, acting on the output of the first, is $y(s) = H(s)\, x(s)$. Therefore, the transfer function of the aggregate system consisting of $H(s)$ and $G(s)$ arranged *in series* is the *product* of the components: $T(s) = H(s)\, G(s) = G(s)\, H(s)$. (Because G and H are merely functions, they commute.)

By similar reasoning, one can show that if two components are arranged in parallel,

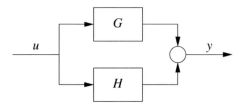

then one can add their respective outputs to find the output of the overall system:

$$y(s) = H(s)u(s) + G(s)u(s) = [H(s) + G(s)]\, u(s)$$

The transfer function of a composite system, consisting of $H(s)$ and $G(s)$ arranged *in parallel* is the *sum* of the components: $T(s) = H(s) + G(s)$.

These two simple rules allow us to handle a variety of open-loop systems. But we also need a rule for closed-loop arrangements. This will lead to a central result in the theory of feedback systems.

The Feedback Equation

Suppose now that we have a desired outcome (a *setpoint*) for the system. In other words, we want the system outcome y to track a given reference signal r as closely as possible. To ensure this behavior, we apply the feedback principle (Chapter 2):

- We compare the actual output y to the reference r.
- We adjust the input to the system to counteract any deviation of y from r.

In other words, if y exceeds r, then we will adjust the input u in such a way that y will be reduced, and vice versa.

Toward this end, we "close the loop" so that the output y can be compared to the setpoint r (see Figure 21-1). All inputs to the circle are summed, and the result is then passed to the controller K. Because the system output y has been multiplied by -1 on its return path, the input to K is the tracking error $e = r - y$. We wish to minimize the magnitude of this error (in other words, we want to reduce it to zero). The controller K transforms the tracking error into a control input u to the system; when the tracking error is zero, no further changes need to be made to the input.

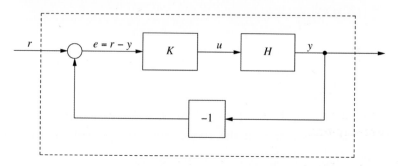

Figure 21-1. The standard feedback loop. The controller K acts on the tracking error e to produce the input u for the plant H. The plant output y is multiplied by –1 on the return path. At the circle, the negative output –y is added to the setpoint r to yield the tracking error e = r – y.

By construction, the system H transforms an input u into an output y. We may now treat the whole assembly as a single system (pictured in Figure 21-1 by the dashed box) that transforms the input r into an output y. What is the correct expression describing the behavior of this *closed-loop* system in terms of its components K and H? It can't simply be the *open-loop* expression $y = HK\, r$, which fails to take into account that the output y is fed back into the system input. Instead, the combined system $H\,K$ acts on the tracking error $e = r - y$ to produce the output y:

$$y = HK(r - y)$$

Because H and K are regular functions, we can solve this equation for y. First multiply out the parentheses on the right-hand side,

$$y = HK(r - y)$$
$$ = HKr - HKy$$

and then bring the second term to the left-hand side:

$$y + H\,Ky = H\,K\,r$$

Now factor out the common factor y on the left:

$$(1 + H\,K)y = H\,K\,r$$

and finally divide by $(1 + HK)$ to obtain

$$y = \frac{HK}{1 + HK}r$$

The transfer function of the closed-loop arrangement shown in Figure 21-1 is therefore

$$T(s) = \frac{H(s)K(s)}{1 + H(s)K(s)}$$

This result is central to all of feedback control theory.

An Alternative Derivation of the Feedback Equation

There is an instructive alternative to deriving the feedback equation, one that makes the iterative aspect of the feedback principle explicit. We begin once again with the basic input/output relation

$$y = HK(r - y)$$

but, instead of solving the equation algebraically, we take an iterative approach by plugging the expression $y = HK\ (r - y)$ back into the equation on the right-hand side:

$$y = HK\left[r - \left(HK(r - y)\right)\right]$$
$$= HKr - (HK)^2 r + (HK)^2 y$$

Now we do it again:

$$y = HKr - (HK)^2 r + (HK)^2 (HK(r - y))$$
$$= HKr - (HK)^2 r + (HK)^3 r - (HK)^3 y$$
$$= \left[HK - (HK)^2 + (HK)^3\right] r - (HK)^3 y$$

and so on. Writing it this way shows more clearly how the output y is being passed through the system HK again and again, and modified each time by the influence of HK. If HK has the effect of *amplifying* its input, then we can see that y will get large very quickly as it goes through the loop repeatedly.

Finally, summing the geometric series in HK will lead us back to the feedback formula. (That's because $x - x^2 + x^3 - \cdots = -x\ (1 - x + x^2 - \cdots) = x/(1 + x)$—provided $|x| < 1$, for otherwise the sum does not converge. This condition provides yet another hint at the constraints that exist for the design of a suitable controller.)

Block-Diagram Algebra

Given the feedback equation that describes a closed-loop arrangement, we now have a set of "rules" that allows us to manipulate block diagrams and find the transfer function of a composite system directly from its block diagram. The rules are summarized in Figure 21-2.

These rules can be used to simplify complex block diagrams and also to modify existing ones. For instance, any series of elements H and G can be replaced by a single element with transfer function HG. Introducing an additional element (such as a filter F) into an existing loop merely amounts to the insertion of an additional factor into the transfer function for the entire loop, and so on.

One can formulate a wide variety of additional rules, but all can be reduced to the three basic rules given here. In any case, the rules shown in Figure 21-2 are sufficient for all block-diagram manipulations in this book.

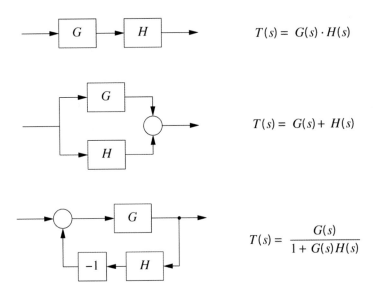

$$T(s) = G(s) \cdot H(s)$$

$$T(s) = G(s) + H(s)$$

$$T(s) = \frac{G(s)}{1 + G(s)H(s)}$$

Figure 21-2. The three most important operations of block-diagram algebra: composition in sequence (top), addition in parallel (center), and the negative feedback loop (bottom).

Limitations and Importance of Transfer Function Methods

The transfer function technology described in the last two chapters may seem like magic: it turns differential equations into simple functions and allows us to manipulate entire control loops through a simple algebra of graphical operations! In the following chapters, we will see how this method allows us also to determine the dynamic response of closed-loop systems from the mathematical structure of the transfer function alone—that is, without actually having to evaluate any time-domain behavior.

That being said, transfer function methods are based on the Laplace transform and are applicable only when certain conditions are met:

- The system dynamics are given by a linear differential equation with constant coefficients.

- Both input and output for all components in the loop are scalars. (In other words, each component has exactly one input and one output.)

There are techniques to extend transfer function methods to more general situations, and there is an alternative formulation of the theory that is not limited to single-input/single-output systems (see Chapter 26).

Beyond the direct applicability of Laplace transforms and transfer functions to performing calculations on specific systems, methods based on Laplace transforms provide a conceptual framework and—in many ways—the terminology for control systems. For this reason alone, it is necessary to gain at least a passing familiarity with them.

CHAPTER 22

PID Controllers

We encountered PID controllers already in Chapter 4. Now we take a closer look at them while using the frequency-space methods introduced in Chapter 20.

The Transfer Function of the PID Controller

As we saw in Chapter 4, the output $u_{\text{PID}}(t)$ of a PID controller in terms of its input $e(t)$ is given by

$$u_{\text{PID}}(t) = k_p e(t) + k_i \int_0^t e(\tau)\, d\tau + k_d \frac{de(t)}{dt}$$

Notice that $u_{\text{PID}}(t)$ is *linear* in $e(t)$. (Both integration and differentiation are linear operations.) The linearity of the PID controller is one reason for its popularity.

We can take the Laplace transform of this expression term by term to obtain the dynamic response of a PID controller in the frequency domain. Using Table 20-1, we find without difficulty that

$$u_{\text{PID}}(s) = \left[k_p + \frac{k_i}{s} + k_d s \right] e(s)$$

The expression within brackets is the transfer function of the controller in frequency space. It has a particularly simple form—in fact, just the factors s and $1/s$ are often used to signify the corresponding terms (see Figure 22-1).

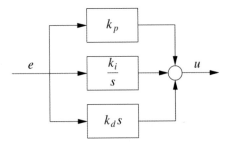

Figure 22-1. The standard form of the three-term or PID controller.

The Canonical Form of the PID Controller

As we have just seen, the transfer function of a PID controller is

$$K(s) = k_p + \frac{k_i}{s} + k_d s$$
$$= \frac{k_p s + k_i + k_d s^2}{s}$$

This is the form most convenient for theoretical work. It has the disadvantage that the three constants (k_p, k_i, and k_d) do not all have the same dimensions, because s has the dimension of frequency, k_i has the dimension of time, and k_d has the dimension of 1/time. In application-oriented contexts, an alternative form of the PID controller transfer function is often used:

$$K(s) = k \left[1 + \frac{1}{sT_i} + sT_d \right]$$

Here k is the controller gain, T_i is the "integral time" (or "reset time"), and T_d is the "derivative time" (or "rate time"). The two forms are equivalent, and the parameters are related:

$$k = k_p \qquad T_i = \frac{k_p}{k_i} \qquad T_d = \frac{k_d}{k_p}$$

Of course, the *numerical values* are different! When comparing values for controller parameters, one must not forget to establish which of the two forms they refer to.

The General Controller

The discussion so far has concerned only controllers that consist of strictly proportional, integral, and derivative terms (PID controllers). This raises the question of what *else* a controller can be. The answer is: anything at all, as long as it has one input and one output and depends only on values that are available at the time that control action is needed. If we want to use Laplace transforms and transfer function technology, then the controller behavior must be describable by a linear and time-invariant differential equation in the time domain. In principle, though, the controller action can be any function of its inputs.

If we were to start entirely from scratch, then what would the most desirable controller look like? Since our intent is for the plant output to track the setpoint signal as closely as possible, the "ideal" controller would act in such a way as to *cancel* the effect of the plant. Under those conditions, the input to the controller r would be precisely the output of the plant y (in an open-loop configuration):

$$y = HK\,r$$

Now assume that K exactly "neutralizes" or cancels the effect of H and so this equation becomes

$$y = r$$

Perfect tracking!

In frequency space, this is easy enough to achieve. Let $H(s)$ be an arbitrary transfer function. Then a controller with the following transfer function will literally *cancel* the effect of the plant H:

$$K_{ideal}(s) = \frac{1}{H(s)}$$

All that remains to do is to transform this transfer function back into the time domain and then build a physical device that exhibits the required dynamical behavior. Of course, there is the rub: controller designs obtained in this way often require arbitrarily large control actions—larger than can be achieved using physical devices. This is not helpful. (But controllers that attempt to cancel the plant approximately are sometimes used as part of a global control strategy; recall the discussion of the Smith predictor in Chapter 11.)

The PID controller, together with a feedback architecture, takes a different approach: the controller does not attempt to cancel the plant dynamics exactly. Instead, it relies on frequent and continuous adjustments in order to have the plant output track the setpoint. For processes with difficult dynamics, however, a controller that is more complicated than a simple three-term controller may lead to better performance.

Proportional Droop Revisited

In Chapter 4, we mentioned the inability of a strictly proportional controller to track a setpoint without incurring a steady-state error. Now, with the controller's transfer function and the feedback equation (Chapter 21) in hand, we can understand this phenomenon more precisely.

For a simple feedback loop with a plant H and a strictly proportional controller $K = k$, the feedback equation is

$$y = \frac{kH}{1+kH} r$$

Now consider the steady state that prevails when all transients have died away. The input r is a constant, and (in the steady state) so is the output of the plant H itself. But under those circumstances, the fraction in the equation just displayed is always less than 1,[1] so that the output from the feedback loop y will always be smaller than the reference value r! As the controller gain k is increased, the fraction will approach 1 and so the steady-state error is reduced. But no finite controller gain will succeed in eliminating it.

---

1. Provided that the controller gain k and the plant output are both positive, as is usually the case.

A Worked Example

Near the end of Chapter 3, we simulated a simple system that reproduced its input but delayed by a single time step. With proportional control, the system could be made to converge to a steady state under certain conditions but always exhibited a noticeable deviation from the setpoint value. We can now calculate the final value that the system converged to.

The system or "plant" H in the example neither increases nor decreases the value of its input; it merely delays it. So as far as the magnitude of the output is concerned, H has no influence and we can replace it with a multiplication by 1. However, the controller K changes the magnitude of its input by a factor of the "controller gain" k. If we again consider only the change in the magnitude, then the feedback formula becomes

$$|y| = \frac{k}{1+k}|r|$$

For the graphs in Figure 3-4, we used values for the controller gain of $k = 0.8$ and $k = 1.1$ while keeping the setpoint constant at $r = 1$. The feedback formula now tells us that the magnitude of the steady-state output should be $\frac{0.8}{1+0.8} \approx 0.44$ and $\frac{1.1}{1+1.1} \approx 0.52$, respectively; these values are also indicated in the figures. (For the unstable case of $k = 1.1$, this result is misleading, of course, because in this scenario the system never settles down to a steady state.)

CHAPTER 23

Poles and Zeros

An important advantage of the "transfer function technology" is that we do not need to evaluate the entire transfer function to obtain information about a system's dynamics. Instead, it is sufficient to merely know the locations of the transfer function's *poles* and *zeros*—that is, the locations where the transfer function diverges or vanishes (respectively) to gain substantial insight into the dynamic behavior.

Structure of a Transfer Function

By construction, the transfer functions for feedback systems tend to be proper rational functions—that is fractions of one polynomial in s over another polynomial in s:

$$H(s) = \frac{N(s)}{D(s)} = \frac{b_m s^m + b_{m-1} s^{m-1} + \cdots + b_0}{s^n + a_{n-1} s^{n-1} + \cdots + a_0}$$

This fact is a consequence of the way transfer functions arise as solutions of linear differential equations with constant coefficients via Laplace transforms. (If the dynamics include time delays, then additional factors of the form e^{-sT} occur in the transfer function—more on this issue later in this chapter.)

The *degree* of a polynomial is the power of its highest term. In the preceding formula, the numerator is of degree m, and the denominator is of degree n. A transfer function is called *strictly proper* if it tends to 0 for large s, and it is called *proper* if it tends to a finite value as s approaches infinity. For transfer functions that are strictly rational functions, this means that for a transfer function to be strictly proper, the degree of the numerator polynomial must be less than the degree of the denominator polynomial ($m < n$) and that the degrees must be equal ($m = n$) for a transfer function to be proper. The *rank R* of a transfer function is the excess of numerator zeros over denominator zeros: $R = m - n$. Transfer functions of physical systems are never improper.

It is always possible to factor a polynomial (in the complex plane). If we factor both the numerator and the denominator polynomials of $H(s)$, we obtain the standard or "root locus" form of the transfer function:

$$H(s) = \frac{k(s - z_1)(s - z_2)\cdots(s - z_m)}{s^r(s - p_1)(s - p_2)\cdots(s - p_{n-r})} \quad \text{with } m \le n$$

If s equals any of the z_k (for $k = 1, ..., m$) then the transfer function vanishes; hence the z_k are the locations of the zeros of $H(s)$, or simply its *zeros*. Similarly, whenever s becomes equal to any of the p_k (for $k = 1, ..., n - r$), the denominator of $H(s)$ vanishes and so the transfer function $H(s)$ "blows up" for that value of s. Such positions are called the *poles* of $H(s)$. Finally, if $r > 0$ then $H(s)$ has a pole of order r at the origin; in that case, one says that $H(s)$ is of *type r*.

Effect of Poles and Zeros

Knowing only the poles and zeros of a transfer function allows us to understand a good deal about the dynamic response of the corresponding system. To see why, we need to understand what poles and zeros can tell us about dynamic behavior.

The effect that zeros have is easy to see: because $y(s) = H(s)u(s)$, the output $y(s)$ is zero whenever $H(s)$ is zero irrespective of the input $u(s)$. Moreover, for values of s near a zero, even though $H(s)$ will not vanish exactly, it will still be small, so that the $y(s)$ will also be small in a neighborhood of a transfer function zero. Zeros block the transmission of signals.

To understand the effect that poles have, we need to work a little harder. We begin with the transfer function in its standard, factored form. Such a rational function can always be split into partial fractions. In other words, we can find coefficients A_k such that

$$H(s) = \frac{A_1}{s - p_1} + \frac{A_2}{s - p_2} + \cdots + \frac{A_n}{s - p_n}$$

To understand how the system responds to a disturbance, we must transform the transfer function back into the time domain. This is now easy to do because we can perform the inverse transformation term by term. According to Table 20-1, an expression of the form $1/(s - p)$ in the frequency domain leads to e^{pt} in the time domain. Each pole p_k contributes an exponential term of the form $e^{p_k t}$ to the transfer function in the time domain. The transfer function in the time domain is therefore a linear combination of exponentials, one for each pole:

$$h(t) = A_1 e^{p_1 t} + A_2 e^{p_2 t} + \cdots + A_n e^{p_n t}$$

Because the transfer function is a representation of the dynamic response of the system, it follows that the dynamic behavior can be expressed as a superposition of "modes." The behavior of each mode is given by the exponential $e^{p_k t}$, where p_k is the position of the corresponding pole.

The dynamic behavior of each mode in the time domain now depends on the value of p_k. If p_k is real and negative, then the dynamic response $e^{p_k t}$ will decay exponentially; the more negative p_k is, the faster is the decay. If p_k is real and positive, then the mode will grow exponentially with time.

But p_k does not need to be real; it can have an imaginary part. In this case, the dynamic response is *oscillatory*. Recall that the exponential function with a purely imaginary argument can be expressed in terms of trigonometric functions (also see Appendix C):

$$e^{i\omega t} = \cos(\omega t) + i \sin(\omega t)$$

The greater ω is, the faster the wiggles.

If the pole is purely imaginary ($p_k = i\omega_k$), then the dynamic response consists of an oscillation with a constant amplitude. If the pole is *complex*—that is, if it has both a real and an imaginary part ($p_k = \sigma_k + i\omega_k$)—then the amplitude of the oscillation will either grow or decay exponentially, depending on the sign of the real part σ_k:

$$e^{p_k t} = e^{(\sigma_k + i\omega_k)t}$$

$$= e^{\sigma_k t}\big(\cos(\omega_k t) + i\sin(\omega_k t)\big)$$

The first term describes the development of the amplitude, and the second term simply wiggles with frequency ω_k. The combination describes an oscillation with time-varying amplitude.

To summarize: the dynamic response is a linear combination of modes, one for each pole. The position of the pole determines the nature of the mode. The most important distinction concerns the sign of the real part of the pole. If the real part is positive, the amplitude of the corresponding mode will grow in time: the system blows up; it is *unstable*. Only poles with a negative real part describe stable behavior. In other words, *for a system to be stable, all of its poles must be in the left-hand plane.* Furthermore, poles with an imaginary part are associated with an oscillatory dynamic response; and the greater (in absolute terms) the imaginary part, the faster the oscillation. (See Figure 23-1.)

Special Cases and Additional Details

The preceding discussion skipped a few details and special cases, which we still need to mention.

Complex conjugate poles

If you consider the time response of a pole with an imaginary part, it may appear as if there will be an *imaginary* time response. After all, if $p_k = \sigma_k + i\omega_k$ then the corresponding mode $e^{\sigma_k t}(\cos(\omega_k t) + i\sin(\omega_k t))$ seems to include an imaginary part (namely, the sine term). But this is not, in fact, the case. By construction, complex poles in the transfer function for a physical system always occur in complex conjugate pairs, so that if $p_k = \sigma_k + i\omega_k$ is a pole, then there will also be a pole $p_{k+1} = \sigma_k - i\omega_k$. Moreover, the corresponding coefficients A_k and A_{k+1} in the partial fraction expansion will also be complex conjugates of each other. Both p_k and p_{k+1} lead to oscillatory modes with the same frequency, but their respective imaginary parts cancel each other out! The end result is that a complex pole, together with its complex conjugate, contributes a purely real oscillatory mode to the dynamic response in the time domain.

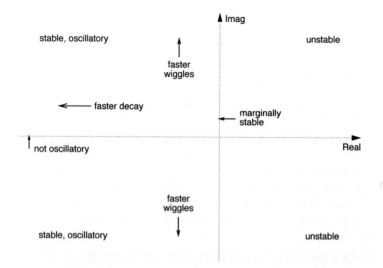

Figure 23-1. Pole positions in the complex plane and how they determine the nature of the dynamic behavior in the time domain.

Multiple poles

It may be that the partial fraction expansion of the transfer function contains terms with the denominator raised to some power:

$$H(s) = \cdots + \frac{A_j}{\left(s - p_j\right)^{\mu}} + \cdots$$

In this case, the corresponding pole is called a *multiple pole of order* μ. According to Table 20-1, the time response for such a term picks up an additional factor of $t^{\mu - 1}$:

$$h(s) = \cdots + A_j t^{\mu-1} e^{p_j t} + \cdots$$

Finally, if a term $A_j/(s - p_j)^{\mu}$ occurs in the partial fraction expansion, then all terms of lower power $(A_k/(s - p_j), A_\ell/(s - p_j)^2, ..., A_m/(s - p_j)^{\mu-1})$ will also occur.

Poles on the imaginary axis

If a pole is purely imaginary (that is, if its real part is zero) then the pole is located on the imaginary axis. The amplitude of the corresponding mode neither increases nor decreases with time, and the mode is said to be *marginally stable*. This is not the case if the pole in question is a multiple pole, because the additional factors of t in the time response for multiple poles make multiple poles on the imaginary axis unstable.

Pole/zero cancellations

It is possible for a pole and a zero to occur at the same location: $p_k = z_j$. In that case, the respective factors in the transfer function will cancel. This effect is sometimes introduced intentionally to suppress an undesired mode, as when additional elements are added to the control loop in order to create a zero at or near the pole of the undesired mode. Even if the cancellation is not perfect, the effect of this operation will be a much reduced amplitude of the mode in question.

Pole Positions and Response Patterns

Figures showing the positions of poles and zeros in the complex plane are known as *pole-zero diagrams*. Poles are indicated using × symbols, and zeros are indicated with open circles. The nature of the system's dynamic behavior can be read off from a pole-zero diagram; Figure 23-3 shows a selection of pole configurations in the complex plane together with the associated dynamic response to a short impulse at $t = 0$. The basic rules are as follows:

- Poles in the right half-plane correspond to modes with amplitudes that grow in time—they are unstable. Amplitudes for poles in the left half-plane diminish in time (they are stable); amplitudes for poles on the imaginary axis remain constant over time (marginally stable). (See panel A in Figure 23-3.)

- Poles on the real axis correspond to modes that are nonoscillatory. Poles off of the real axis correspond to modes that are oscillatory; such poles always occur in complex-conjugate pairs. (See panel B in Figure 23-3.)

- A multiple pole on the imaginary axis is unstable. The amplitude of the associated mode grows only as a power law (not exponentially). (See panel C in Figure 23-3; the pole at the origin is a double pole, indicated by the "2" in the pole-zero diagram, and its dynamic response increases linearly with time.)

- Moving poles *vertically* away from the real axis increases the frequency of the corresponding, oscillatory mode. (See panels B and C in Figure 23-3.)

- Moving stable poles *horizontally* away from the imaginary axis makes the amplitude change faster: moving a stable pole to the left makes the corresponding mode decay faster, whereas moving an unstable pole to the right makes the amplitude increase faster. (See panels B and D in Figure 23-3.)

- A *zero* near a pole diminishes that pole's importance by reducing the amplitude of the corresponding mode.

- Zeros in the right half-plane are indicative of non-minimum phase systems. The initial response of such systems will be in the *opposite* direction of the input.

Dominant Poles

If we are dealing with a transfer function of high order (many poles and zeros), then we can often simplify the analysis by concentrating only on the *dominant poles* and neglecting the other ones. Poles that can be neglected are those that are far to the left of the dominant poles: the dynamic response of the neglected poles will decay quickly, so they will not exert a major influence on system behavior. We can also neglect poles that are close to a zero; the amplitude of the associated mode will be very small and, again, ignoring such a pole won't change the observed behavior much. (See Figure 23-2.)

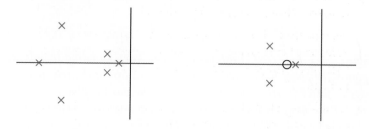

Figure 23-2. Dominant poles in red and subdominant poles in green. In the right-hand graph, the pole on the real axis is subdominant because it is nearly canceled by the nearby zero.

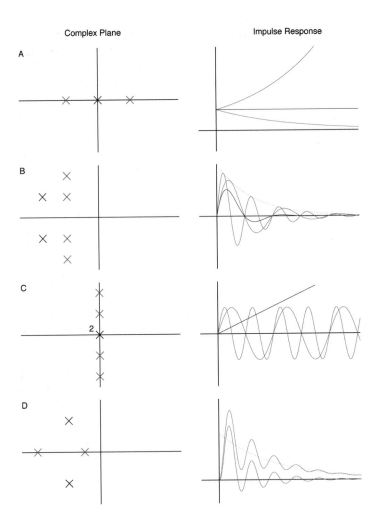

Figure 23-3. Pole configurations and their corresponding impulse responses. In panel B, the green curve decays as fast as the red curve but wiggles faster (vertical pole movement), whereas the blue curve oscillates at the same frequency as the red one but decays faster (horizontal pole movement). In panel C, the blue pole at the origin is a double pole (indicated by the "2"), and its dynamic response grows linearly with time. In panel D, the curves correspond to the blue pair of poles and either the red or the green pole. The green pole is far from the origin and hardly affects the dynamic response, but the red pole is close to the origin and therefore determines the dominant, slow decay of the red curve.

The time response in panel D of Figure 23-3 consists of the contribution of the blue poles and either the red or the green pole. The blue pair of complex conjugate poles describes a damped oscillation. The green pole, which is farther away from the origin than the blue poles, describes a mode that decays so quickly that it hardly affects the dynamic response at all. The red pole, however, is closer to the origin than the blue poles. Its corresponding mode in the time domain decays more slowly than the oscillations due to the complex conjugate pair. The red pole is a typical dominant or "slow" pole.

Pole Placement

Knowing how the dynamic response of a system is determined by the position of the system's poles, we can attempt to design a system with the desired behavior by "moving its poles"—a process known as *pole placement*. Controller tuning (see Chapter 9) is a simple example of pole placement: we try to adjust the controller gains in order to "move the poles" into positions corresponding to the desired performance. The root locus method (Chapter 24) is a graphical design technique based on the idea of moving poles in the complex plane.

At the beginning of Chapter 9 we listed several goals for the tuning process, such as the typical response time of the system, the amount of overshoot after a disturbance, the damping of oscillations, and so on. These quantities can be related directly to pole positions.[1] Specifying numerical values for the tuning goals therefore amounts to restricting the system's poles to a specific region of the complex plane. Controller tuning can now be regarded as a process of moving the dominant poles into the permissible regions.

But we are not limited to adjusting the controller gains. One can insert additional elements with known transfer functions into the control loop, a process known as "loop shaping." For instance, one can introduce a "lead/lag compensator" with transfer function

$$G(s) = \frac{1 + as}{1 + bs}$$

into the control loop. This compensator has a pole at $-1/b$ and a zero at $-1/a$. We can now adjust a such that the zero coincides (for example)

---

1. See for example *The Art of Control Engineering* by K. Dutton, et al. (1997).

with an undesirable pole and fix b to place the corresponding pole into favorable position.

What to Do About Delays

All the preceding considerations assume that the transfer function is, in fact, a purely rational function: one polynomial in s divided by another polynomial in s. If the dynamics of the system can be expressed entirely in terms of a linear differential equation with constant coefficients, then this assumption will be satisfied. But it will *not* be satisfied if the system dynamics include *pure time delays*.

In physical systems, delays often arise because energy or material is transported over some distance: if water needs to flow through a hose or pipe, then there will be a delay from the time the faucet was opened to the time the water level in the bucket begins to rise. Symbolically, the output of a system H is delayed by some time τ relative to the input

$$y(t) = H\,u(t - \tau)$$

Time delays are not described by linear differential equations with constant coefficients, and their transfer functions are not purely rational functions—instead, time delays introduce exponential factors of the form $e^{-s\tau}$ into the transfer function. This is less of a mathematical problem (the transfer functions remain well behaved) but it is a major *practical* inconvenience. Many analytical methods in control theory rely on the transfer function being the ratio of two polynomials. (For instance, the ability to perform a partial fraction expansion—which was central to our description of the dynamic response in terms of modes and their associated poles—requires the transfer function to be a rational function.)

In order to retain the desired structure of the transfer function, one approximates the exponential factor $e^{-s\tau}$ either by a power series or by a rational function approximation. All of these approximations are good as long as $s\tau$ is small—that is, for time scales that are long compared to the duration of the delay.

The Taylor expansion of the exponential function is familiar, but it is not a good approximation unless $s\tau$ is quite small:

$$e^{-\tau s} \approx 1 - \tau s + \frac{(\tau s)^2}{2} \mp \cdots$$

Much better results can be obtained by approximating the exponential function through a rational function instead of a power series (Padé approximation). The following two approximations are often used:

$$e^{-\tau s} \approx \frac{1 - \tau s / 2}{1 + \tau s / 2}$$

$$\approx \frac{2 - \tau s + (\tau s)^2 / 6}{2 + \tau s + (\tau s)^2 / 6}$$

Finally, we can make use of the identity $\lim_{n \to \infty} (1 + x/n)^n = e^x$, and thereby arrive at an approximation for the exponential function, by plugging in some "large," but finite value of n (say, $n = 5, ..., 20$):

$$e^{-\tau s} \approx \frac{1}{(1 + \tau s / n)^n} \qquad n \gg 1$$

This approximation has a nice interpretation as a sequence of n lags, each with time constant τ/n. The value of n should not be too large, because otherwise the expression becomes numerically unstable.

Finally, keep in mind that all of these formulas are only *approximations*. Not only do they have a limited range of validity, but using them does also change the pole structure of the transfer function. The results should therefore be used with care!

CHAPTER 24

Root Locus Techniques

As we saw in Chapter 23, the location of the poles and zeros of the transfer function determines the system's dynamic behavior. We can therefore change the dynamics of the system by moving the poles and zeros to more desirable positions, a method known as "pole placement." The easiest way to do this is by adjusting the controller gains—that is by "tuning" the controller. (See Chapter 9 for more hands-on techniques of controller tuning.)

As the controller gains are varied, the poles and zeros of the closed-loop transfer function trace out curves in the complex plane that are called *root locus curves*. A root locus diagram is a plot of the complex plane showing the root locus curves[1] as the gain is increased from zero toward infinity. Given such a diagram, we can choose the gain value that moves the dominant poles closest to their desired locations.

Because the structure of transfer functions is not arbitrary (they tend to be rational polynomials), we can make some general statements about global features of the corresponding root locus diagrams. These rules are discussed next.

---

1. A note on terminology: I consider a root locus to be the position of a single solution of the characteristic equation. The root locus curve is the set of all such positions as the gain is varied.

Construction of Root Locus Diagrams

Root locus diagrams are usually drawn for closed-loop systems, such as the one depicted in Figure 24-1. This system has the closed-loop transfer function

$$T(s) = \frac{KG}{1+KGH}$$

This function has a pole when the denominator becomes zero, so the condition for a pole is

$$1 + KGH = 0$$

This equation is also called the "characteristic equation" of the system. Note that the characteristic equation of the *closed-loop* system involves only the *open-loop* transfer function $K\,G\,H$.

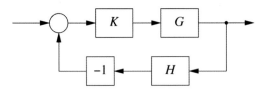

Figure 24-1. A closed-loop arrangement, involving controller K, plant G, and return filter H.

Now consider the case of purely proportional control. In that case, the controller becomes $K(s) = k$, where k is the controller gain. With this choice of controller, the open-loop transfer function simplifies to kGH.

Transfer functions tend to be rational functions. Let us write out the numerator and denominator of the open-loop transfer function kGH explicitly while factoring out the scalar gain k:

$$kGH = k\frac{N}{D}$$

where $N(s)$ is the numerator of the open-loop transfer function and $D(s)$ is the denominator. Multiplying through by D, the characteristic equation can now be written as

---

$$D(s) + kN(s) = 0$$

Recall that those values s for which this equation is satisfied are the poles of the *closed-loop* transfer function and that all such values of s make up the root locus curves.

Now consider the two limiting cases of $k \to 0$ and $k \to \infty$. For $k = 0$, the characteristic equation reduces to $D(s) = 0$. In other words, in the limit of $k = 0$, the *closed-loop* transfer functions has poles when the denominator of the *open-loop* transfer function is zero: the poles of the open-loop and the closed-loop transfer functions coincide for $k = 0$. For the other limit, first divide through by k to obtain $D(s)/k + N(s) = 0$ and then let $k \to \infty$; we are left with $N(s) = 0$. In this limit, the poles of the *closed-loop* transfer function coincide with the zeros of the *open-loop* transfer function (that is, those values of s for which the numerator $N(s)$ of the open-loop transfer function vanishes.)

These observations lead to the following conclusion: *the root locus curves begin at the poles of the open-loop transfer function for $k = 0$ and approach the zeros of the open-loop transfer function for $k \to \infty$.*

A more detailed description is given in the next section.

Root Locus or "Evans" Rules

Assume that the complete open-loop transfer function can be factored in the following way:

$$K(s)\, G(s)\, H(s) = k \cdot \frac{(s-z_1)(s-z_2)\cdots(s-z_m)}{(s-p_1)(s-p_2)\cdots(s-p_n)} \quad \text{with } k \geq 0 \text{ and } n \geq m$$

The transfer function has n poles and m zeros, and the condition $n \geq m$ ensures that it is proper.

We can now state[2] the following rules about the global appearance of a root locus diagram for nonnegative values of k. (These rules are also known as "Evans rules", after W. R. Evans, who first formulated them.)

---

2. Derivations can be found, for example, in *Modern Control Engineering* by K. Ogata (2009).

1. The root locus diagram is symmetrical with respect to the real axis.

2. There are n branches in the root locus diagram.

3. Every pole is a starting point ($k = 0$) of a branch. All branches begin at a pole.

4. Every zero is an endpoint ($k \to \infty$) of a branch.

5. If there is an excess of poles over zeros, then the remaining $R = n - m$ branches tend to infinity as k becomes large. (These branches end at the "zeros at infinity" of the transfer function.)

6. If the transfer function has R more poles than zeros, then the corresponding branches are asymptotic to straight lines as $k \to \infty$. The asymptotes are at the following angles with the positive real axis:

$$\frac{(2\nu + 1)\pi}{R} \qquad \nu = 0, 1, 2, \ldots, R - 1$$

7. All asymptotes intersect in a point on the real axis. The position σ_0 of this intersection point is given by

$$\sigma_0 = \frac{1}{R}\left(\sum_{j=1}^{n} p_j - \sum_{j=1}^{m} z_j \right)$$

8. If the transfer function has poles or zeros on the real axis, then those sections of the real axis that have an odd number of poles and zeros to their right will be part of one of the branches. Multiple poles or zeros are counted multiple times. (See examples later in this chapter!)

9. Different branches may intersect each other. Such points of intersection are called "singular points." Singular points are values of s that satisfy the following equality:[3]

$$\sum_{j=1}^{n} \frac{1}{s - p_j} = \sum_{j=1}^{m} \frac{1}{s - z_j}$$

---

3. The general condition for a singular point is that $dk/ds = 0$, where $k = -D(s)/N(s)$. If the transfer function factors completely, then this condition yields the formula given in the text.

This is a necessary condition only—there may be solutions of this equation that correspond to negative values of k.

10. Branches intersecting on the real axis will depart from (or arrive on) the real axis at right angles to the real axis.

Employing these rules, a root locus diagram can be sketched by following this sequence of steps:

1. Plot the positions of the poles and zeros of the open-loop transfer function in the complex plane.

2. Identify those sections of the real axis that are part of one of the branches.

3. Plot the position where the asymptotes intersect.

4. Plot the asymptotes.

5. Plot the critical points (intersections of branches).

6. Fill in the missing parts of the branches.

Angle and Magnitude Criteria

We can use the polar representation of complex numbers to derive two interesting conditions for a point s to lie on a root locus curve. Every complex number can be written in the polar form $z = |z|\, e^{i\phi}$, where $\phi = \arg z$ is the phase angle of z. Assume that the transfer function splits into factors as defined previously. We now use the polar representation of each factor $(s - p_j)$ and $(s - z_j)$:

$$(s - p_j) = |s - p_j|\, e^{i\, \arg(s-p_j)} \quad \text{and} \quad (s - z_j) = |s - z_j|\, e^{i\, \arg(s-z_j)}$$

to write the transfer function as

$$K(s)\, G(s)\, H(s) = k \cdot \frac{|s - z_1|\, |s - z_2| \cdots |s - z_m|\, e^{i[(\arg(s-z_1)+\arg(s-z_2)+\cdots+\arg(s-z_m)]}}{|s - p_1|\, |s - p_2| \cdots |s - p_n|\, e^{i[(\arg(s-p_1)+\arg(s-p_2)+\cdots+\arg(s-p_n)]}}$$

The characteristic equation requires that $K(s)G(s)H(s) = -1$. Using the polar representation of the transfer function just derived, we can write the characteristic equation as two separate equations for the magnitude and the phase angle. This leads to two conditions: the *magnitude condition,*

$$k \cdot \frac{|s - z_1|\, |s - z_2| \cdots |s - z_m|}{|s - p_1|\, |s - p_2| \cdots |s - p_n|} = 1$$

and the *angle* condition,

$$\sum_{j=1}^{m} \arg(s - z_j) - \sum_{j=1}^{n} \arg(s - p_j) = (2\nu - 1)\pi \quad \nu = 1, 2, 3, \ldots$$

Because they must be satisfied simultaneously by any point s on a root locus curve, these two conditions can be used to determine whether a point belongs to one of the curves.[4]

Practical Issues

Root locus diagrams are a great way to develop a sense for the overall dynamic behavior of a system—although still fairly abstract, the map of roots and zeros provides more intuition than can a formula for the transfer function! At the same time, it is worth remembering that it is the dominant poles (those closest to the origin) that determine the behavior of the system. The global structure of the diagram is of less interest than the area near the origin.

The root locus diagram is an analytical technique. It requires an analytic expression for the transfer function of the plant or process, and it becomes more useful as the transfer functions become more complicated. If there is no good theoretical model of the system and one must employ a phenomenological descriptions obtained from experiments (see Chapter 8), then root locus diagrams are less necessary. In fact, as long as the transfer function is simple enough, one can work out the optimal positions of its dominant poles and zeros ahead of time and obtain "plug-in" rules for the controller gains (this is how the semi-analytical tuning methods in Chapter 9 work).

The root locus diagram is limited to displaying the movements of the poles and zeros as a single scalar parameter is varied. That is frequently not enough, since using a PI controller means that there are already *two* gain parameters to worry about. Employing a three-term controller or introducing a smoothing filter adds further adjustable parameters. Ultimately, this means creating a sequence of root locus di-

---

4. They can also be used to obtain further information about the appearance of a root locus diagram, such as the angles under which root locus curves enter or leave a pole or zero. For more information, see *The Art of Control Engineering* by K. Dutton, et al. (1997).

agrams, one for each value of the secondary parameter. (We'll see an example later in this chapter.)

In most cases, root locus diagrams will be drawn with the aid of a computer. Specialized plotting programs exist[5] that take into account the special structure of the diagram and utilize the Evans rules to generate the plot. These programs usually require that the transfer function be specified explicitly as a strictly rational function.

Examples

In Chapter 20, we derived the transfer function for a simple lag:

$$G(s) = \frac{1}{1 + sT}$$

This function describes systems, such as a heated vessel, that exhibit a particularly simple dynamic. If such a system undergoes a steplike change of its input, then its output will slowly approach its new steady-state value without oscillation or overshooting. (The temperature in a heated vessel increases steadily if the heat is turned on, and it decreases to the ambient temperature again if the heat is turned off.) Because the output does not respond immediately to the change in input, such systems are referred to as "simple lags." The response to a step input of magnitude C that occurs at $t = 0$ is given by an exponential function in the time domain:

$$y(s) = C\left(1 - e^{-t/T}\right)$$

We have encountered this formula already as an approximate process model when looking for a phenomenological description of a system's dynamic response (see Chapter 8 and Chapter 9).

Simple Lag with a P Controller

Consider a simple-lag system in combination with a proportional controller but without a return filter. In this case, we have

---

5. Matlab, Scilab, and Octave all include routines for creating root locus diagrams.

$$K(s) = k$$

$$G(s) = \frac{1}{1 + sT}$$

$$H(s) = 1$$

The open-loop transfer function is therefore

$$K(s)\, G(s)\, H(s) = \frac{k}{1 + sT}$$

and the characteristic equation is

$$\frac{k}{1 + sT} = -1$$

The open-loop transfer function has one pole (at $s_0 = -1/T$) and no zeros, so we expect *one* asymptote at angle π with the positive real axis. In other words, the asymptote is parallel to the negative real axis.

In order to create a figure, we must assign a numerical value to the parameter T. Let's set $T = 1$, which means that we measure time in units of T (see Figure 24-2).

The root locus diagram agrees with our expectations: there is only a single branch, it is aligned with the negative real axis, and it exhibits no oscillatory behavior. Increasing the controller gain k moves the pole to the left, starting from the pole at $s_0 = -1$ and moving toward the "zero at infinity." The part of the real axis to the left of the pole has "an odd number of poles (namely one) to its right" and is therefore part of the root locus curve (compare Rule 8).

A general problem with root locus diagrams is that the controller gain k is not shown. In Figure 24-2, we indicated several distinct values of k with symbols in the root locus diagram (top panel) and showed the corresponding step responses in the bottom panel. It is customary to use arrows to indicate the direction of increasing controller gain on each branch of a root locus diagram.

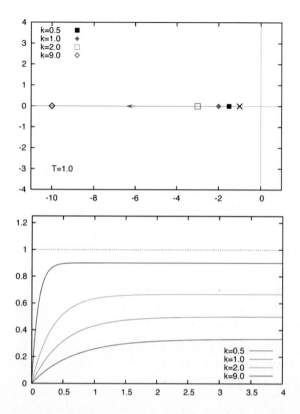

Figure 24-2. Root locus diagram (top) and step response (bottom) for a simple lag under strictly proportional control. The pole of the open-loop transfer function is indicated by a × sign. Also indicated are the pole positions for various gain values; the corresponding step responses are shown in the bottom panel.

Simple Lag with a PI Controller

As discussed in Chapter 4, proportional control is generally insufficient to give good tracking performance because it will lead to "proportional droop." This is evident in the step responses shown in Figure 24-2: even though increasing the gain increases the response time and reduces the tracking error, none of the curves manage to reach the input value in the steady state.

We therefore include an integral term in the controller, so that its transfer function now becomes

$$K(s) = k_p + \frac{k_i}{s}$$

$$= k \left(1 + \frac{1}{sT_i} \right)$$

$$= k \left(\frac{sT_i + 1}{sT_i} \right)$$

To construct a root locus diagram, we must use the alternate representation of the controller transfer function that involves the integral time constant T_i. Only this form of the transfer function allows us to factor out the variable parameter k, as is required.

For a PI controller and the simple-lag system considered previously, the characteristic equation is thus

$$k \left(1 + \frac{1}{sT_i} \right) \left(\frac{1}{1 + sT} \right) = -1$$

This system has one zero (at $z_0 = -1/T_i$) and two poles: one at the origin ($p_0 = 0$) and the other at $p_1 = -1/T$. Because there are two poles, we expect two branches; however, because the excess of poles over zeros is still $R = 1$, there is only a single asymptote, which is aligned with the negative real axis. The behavior in the time domain is a combination of two distinct contributions, one from each branch.

We should expect the diagram to look topologically different when the zero is between the two poles than when it is to the left of both—in other words, depending on whether (respectively) $T_i > T$ or $T_i < T$. (This is a good example for the kind of consideration one must undertake when dealing with more than a single varying parameter in a root locus plot.) Figure 24-3 and Figure 24-4 show both cases.

If $T_i > T$ then the zero lies between the two poles. The root locus curves are entirely real, and the step-response behavior is monotonic without oscillations. Observe again how the root locus curves include only those parts of the real axis with an odd number of poles and zeros to their right.

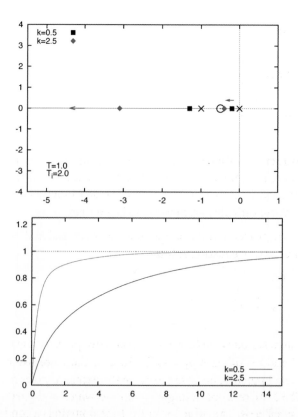

Figure 24-3. Root locus diagram (top) and step response (bottom) for a simple lag under PI control, in a configuration where the controller's integral time T_i is greater than the system's time constant T: $T_i >$ T. The poles of the open-loop transfer function are indicated using $\times$ signs, the zero is marked by a circle. (See also the caption to Figure 24-2.)

The step response for the lower value of the controller gain ($k = 0.5$) is dominated by the "slow" pole at -0.2, but in the step response for the higher value of the controller gain ($k = 2.5$) we can nicely see the effect of both poles. In this case, the amplitude of the slow pole at -0.4 is strongly reduced by its vicinity to the zero at -0.5, and so the initial response is dominated by the fast pole at -3.1. Only after the fast behavior has decayed does the response of the slow mode become apparent (and with a small amplitude).

Finally, if $T_i < T$ then the root locus curves are no longer entirely confined to the real axis. When using values of k for which the poles have acquired an imaginary part, we find oscillatory behavior in the time domain. For this system, we find oscillatory behavior only for an intermediate range of controller gains: further increases in the controller gain lead again to nonoscillatory behavior, albeit with an initial overshoot.

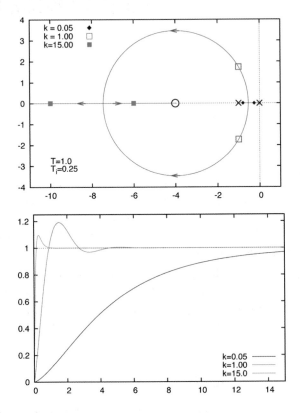

Figure 24-4. Root locus diagram (top) and step response (bottom) for a simple lag under PI control, in a configuration where the controller's integral time T_i is smaller than the system's time constant T: $T_i < T$. The poles of the open-loop transfer function are indicated using × signs, the zero is marked by a circle. (See also the caption to Figure 24-2.)

CHAPTER 25

Frequency Response and the Bode Plot

In this chapter we study the response of a system subject to an oscillatory input. In particular, we will ask how the output changes when the frequency of the input signal is varied. We will also introduce the Bode plot, which is a versatile method of representing a system's frequency response graphically.

The topics treated in this chapter are the starting point for many forms of more advanced analysis. For the most part, they rely on having detailed knowledge of a system's transfer function. The Bode plot, however, is a pretty straightforward technique that is quite generally useful.

Frequency Response

When trying to understand the dynamic response of a system, it is often useful to study how the system responds to *sinusoidal* input signals of differing frequency. Such signals are (of course) the natural description for any form of oscillatory behavior. Furthermore, because the inverse of the frequency $\omega = 2\pi/T$ defines a time scale, the response of a system to a sinusoidal input with frequency ω provides information about the response to a more general disturbance that occurs on a time scale comparable to T.

Frequency Response in the Physical World

There is a very general pattern for the dynamic response of objects in the physical world as a function of the stimulating frequency. At very low frequencies, the object will follow the input faithfully and with only a small delay. (That is, small when compared to the period T of the input signal.) The reason for this behavior is clear: as long as the input changes sufficiently slowly, the system has enough time to adjust and can therefore replicate the input signal.

As we go to the opposite limit of very high frequency, systems will have an increasingly hard time to "keep up" with the input signal; after all, objects in the physical world can move only so fast. As a result, the amplitude of the dynamic response shrinks toward zero for very high frequencies. Moreover, the output will be phase shifted and will lag behind the input. This is a plausible result: the system was assumed to be linear, and all a linear system can do is to shift and scale its input.

In this way, the behavior of pretty much *any* real-world system is fixed in the two limits of very high and very low frequency inputs. Between these two extremes lies the entire range of possible behaviors. In particular, systems that have a tendency to oscillate at a certain frequency will easily be excited by inputs whose frequency is close to their natural (or resonance) frequency. Complex assemblies may have several such resonance frequencies and exhibit a complicated dynamic response as the input frequency is varied.

The frequency response is not just a conceptual construct: it can be observed directly. One applies a sinusoidal input signal, waits until all transient behavior has disappeared, and then measures the amplitude and phase of the output (relative to the input). This procedure is repeated over the entire range of frequencies of interest. One can even buy devices that perform the entire program automatically; such devices are known as *spectrum analyzers*.

Frequency Response for Transfer Functions

If a sinusoidal signal of frequency ω is passed to a linear system $G(s)$ as input, then the resulting output is also a sine wave with the same frequency ω, but the amplitude M and phase ϕ of the output will be altered compared to the input. This is a plausible result: the system was assumed to be linear, and all a linear system can do is to shift and scale its input.

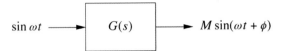

$$\sin \omega t \longrightarrow \boxed{G(s)} \longrightarrow M \sin(\omega t + \phi)$$

Possibly more surprising is that both amplitude and phase of the output signal can be extracted easily from the transfer function $G(s)$ when it is written in polar coordinates and evaluated for the purely imaginary frequency iω. Because $G(s)$ is a complex-valued function, we can write it in polar coordinates as

$$G(s) = M(s)e^{i\phi(s)}$$

We now evaluate $G(s)$ for the purely imaginary frequency $s = i\omega$, where ω is the frequency of the input signal. In this limit, amplitude and phase of the system output are given by the magnitude and phase of the transfer function:[1]

$$M(\omega) = |G(i\omega)| \qquad \text{Amplitude}$$
$$\phi(\omega) = \arg G(i\omega) \qquad \text{Phase shift}$$

Observe that both the amplitude and the phase shift are functions of the input frequency ω.

A Worked Example

To develop a sense for the type of manipulation involved, let's consider the simple lag. It has the straightforward transfer function

$$G(s) = \frac{1}{1 + sT}$$

which we now evaluate for $s \to i\omega$:

---

1. This is not obvious. One begins with the transfer function acting on the sinusoidal input: $G(s)\frac{\omega}{s^2 + \omega^2}$, where we have used the representation of the sine function in the frequency domain. To find the output signal, this expression is transformed back into the time domain using partial fractions. The output signal turns out to be a scaled and shifted sine. In the course of the partial-fraction expansion, one is led to evaluate the transfer function in the limit of $s \to i\omega$.

$$G(i\omega) = \frac{1}{1+i\omega T}$$

The denominator is a complex number. To make it real, we multiply both numerator and denominator by the denominator's complex conjugate (see Appendix C for a refresher on the manipulation of complex variables):

$$G(i\omega) = \frac{1}{1+i\omega T}\frac{1-i\omega T}{1-i\omega T}$$

$$= \frac{1-i\omega T}{1+\omega^2 T^2}$$

$$= \frac{1}{1+\omega^2 T^2} - i\frac{\omega T}{1+\omega^2 T^2}$$

We can now easily obtain the phase shift as

$$\phi(\omega) = \arctan\frac{\text{imaginary part}}{\text{real part}}$$

$$= \arctan\frac{-\omega T}{1}$$

$$= -\arctan\omega T$$

To find the magnitude, we can proceed as follows:

$$M(\omega) = |G(i\omega)| = \left|\frac{1}{1+i\omega T}\right|$$

$$= \frac{|1|}{|1+i\omega T|}$$

$$= \frac{1}{\sqrt{(1+i\omega T)(1-i\omega T)}}$$

$$= \frac{1}{\sqrt{1+\omega^2 T^2}}$$

It is clear that the algebra quickly becomes formidable, especially when we are considering more complicated transfer functions. There are many labor-saving tricks for hand calculations, but if all we need is a graph (such as a Bode plot) then a plotting program that can handle complex numbers may be sufficient.[2]

The Bode Plot

The Bode plot shows both amplitude $M(\omega)$ and phase shift $\phi(\omega)$ as functions of the input frequency ω, using logarithmic scales for the amplitude and frequency. Both curves can be combined in a single graph, but it is often more convenient to use different panels for amplitude and phase.

Figure 25-1 shows the step response for two systems (top panel), and the corresponding Bode plots (bottom panel). The transfer functions for the two systems are as follows:

$$H(s) = \frac{1}{1 + sT} \qquad \text{Simple lag}$$

$$H(s) = \frac{\omega_0^2}{s^2 + 2\zeta\omega_0 s + \omega_0^2} \qquad \text{Damped harmonic oscillator}$$

When examining the Bode plot, we observe first of all that our previous considerations regarding the frequency response of physical systems are confirmed. Both systems follow the input faithfully for low frequencies but lag behind with diminishing amplitude in the high-frequency limit.

Beyond these general observations, there are some specific details that stand out. Most notable is the resonant peak in the amplitude of the oscillator, which occurs when the input frequency equals the oscillator's natural frequency. Whereas the location of the peak is determined by the resonant frequency, the height of the peak depends on the amount of damping: the lower the damping, the higher the peak. If the damping is sufficiently strong, then oscillations are completely suppressed. In this case, the peak will disappear entirely. (Think of a mass/spring system in which the mass is embedded in honey.)

The amplitude diminishes for high frequencies in a manner that depends on the asymptotic behavior of the transfer function in this limit: the transfer function for the simple lag decays as $1/\omega$, whereas the transfer function for the oscillator decays as $1/\omega^2$. This distinction can

---

2. Gnuplot and the major calculational packages (Matlab, Scilab, Octave, ...) can operate with complex quantities directly.

be seen in the different slopes of the amplitude for large ω. Since the amplitude panel of a Bode plot is simply a double logarithmic plot, power-law behavior like this shows up as straight lines.

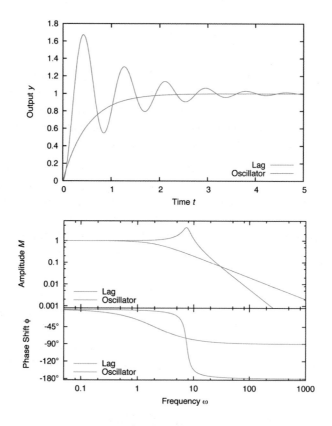

Figure 25-1. Step response (top) and Bode plot (bottom) for both a simple lag and an oscillator.

Similar considerations apply to the phase shift. In the high-frequency limit, the phase is determined by the asymptotically dominant term in the transfer function. For the simple lag, this is

$$G(i\omega) = \frac{1}{1 + i\omega T} \approx \frac{1}{i\omega T} \qquad \text{for large } \omega$$

$$= \frac{-i}{\omega T} \qquad \text{multiplying with } \frac{-i}{-i}$$

$$= \frac{1}{\omega T} e^{-i\pi/2} \qquad \text{since } -i = e^{-i\pi/2}$$

The last line is in the form of the polar representation of a complex number: $z = r\, e^{i\phi}$. We can therefore identify the phase as $\phi = -\pi/2$ (= −90 degrees) in the high-frequency limit, which agrees with Figure 25-1. An analogous consideration for the harmonic oscillator leads to a phase lag of $\phi = -\pi$ (= −180 degrees).

We have explained the appearance of the Bode plot in terms of the transfer function's form, but one commonly argues the other way around. From features of the Bode plot (such as asymptotic behavior or the presence and location of resonant peaks) one can determine what terms must be present in the transfer function. A numerical fit can then serve to fix the outstanding numerical parameters (like T, ζ, or ω_0 in our examples).

One can even go a step further and ask what additional elements with known transfer function (such as lags or filters) should be *added* into the control loop to change the Bode plot—and, by implication, the system behavior—in some desirable fashion. This process is known as *loop shaping* and is an important topic in the design of electronic circuits.

A Criterion for Marginal Stability

We learned in Chapter 23 that a closed-loop system as shown in Figure 25-2 is *marginally stable* if its transfer function

$$T(s) = \frac{G(s)}{1 + G(s)}$$

has poles on the imaginary axis. The closed-loop transfer function has a pole when its denominator vanishes. The condition for marginal stability therefore becomes

$$1 + G(i\omega) = 0 \qquad \text{or} \qquad G(i\omega) = -1$$

where ω is a real-valued frequency (so that iω is a point on the imaginary axis).

In this chapter we have seen that the frequency response of the system $G(s)$ is given by its transfer function evaluated along the imaginary axis.

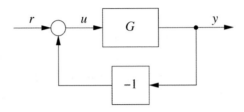

Figure 25-2. A closed-loop arrangement.

We can combine these two observations and express the criterion for marginal stability of the closed-loop system in terms of the frequency response of the open-loop system: *a closed-loop system is marginally stable if the frequency response G(iω) of the open-loop system equals –1 for any frequency ω.*

For any fixed s, we know that $G(s)$ is simply a complex number that can be written in polar form. Hence the closed-loop system is marginally stable when

$$G(i\omega) = r(\omega)\, e^{i\phi(\omega)} = -1$$

or, equivalently, when

$$r(\omega) = 1$$
$$\phi(\omega) = -\pi$$

In other words, the closed-loop system is marginally stable when the open-loop system introduces a phase lag of half a period ($\phi = -\pi$) while at the same time maintaining the amplitude of the input signal.

This makes imminent sense. The phase lag of half a period means that the return signal is perfectly out of phase with the input, but the multiplication by –1 on the return paths means that the return signal ends up being in phase with the input signal.

Under the conditions just derived, the closed-loop system is unstable when supplied with a nonzero input. It is amusing to note that if the input is zero ($r = 0$ in Figure 25-2) then the system will exhibit self-sustained oscillations: since there is perfect constructive feedback but no damping, any oscillation in the loop will persist forever.

For a stable system, in contrast, either one of the following two conditions must hold.

- At the *gain crossover frequency*, where the open-loop gain is 1, the open-loop phase shift must be less than $-\pi$ ($= -180$ degrees).

- At the *phase crossover frequency*, where the open-loop phase shift is $-\pi$ ($= -180$ degrees), the open-loop gain must be less than 1.

These conditions simply give an upper bound on gain and phase. To ensure safe operations, we want the system to be far enough away from the upper bounds. This can be expressed through the concept of a *stability margin*:

- *Phase margin* is the angle by which the phase shift is smaller than $-\pi$ at the gain crossover frequency.

- *Gain margin* is the factor by which the gain is smaller than 1 at the phase crossover frequency.

As a rule of thumb, the gain margin should be between 2.0 and 2.5 and the phase margin should be between $\pi/4$ and $\pi/3$.

Other Graphical Techniques

In addition to the Bode plot (and the root locus diagram; see Chapter 24) other graphical techniques have been developed to visualize the frequency response for feedback systems. The two most important of them are the *Nyquist plot* and the *Nichols chart*.

The frequency response consists of two quantities: amplitude and phase. In the Bode plot, both of them are plotted separately as functions of the input frequency. This representation brings out the dependency on ω, but it requires two separate panels. In the Nyquist plot, the two quantities M and ϕ are plotted in a single, two-dimensional plot, using polar coordinates: the amplitude M is plotted as the radius of the angle ϕ, while ω assumes all positive values. The Nyquist plot shows both quantities in a single graph, but the explicit dependence on the input frequency is lost.

The Nichols chart is also a two-dimensional graph, but it uses rectangular coordinates rather than polar coordinates. In the Nichols chart, the amplitude is plotted as a function of the phase shift while ω again runs through all positive values.

Various rules exist to identify properties of the system from these graphs by examining the position and shape of the curve describing the system relative to various special locations on the graph. The Nyquist plot specifically is also a starting point for deeper analytical studies that bring the machinery of complex function theory to bear on these problems.

These techniques tend to become quite specific to problems arising for systems that must operate over a wide range of frequencies—for example, electrical amplifiers used in telecommunications. Because they seem less applicable to the types of questions of greatest concern to us, we won't pursue the topic here.[3]

3. For a first introduction, see *Schaum's Outline of Feedback and Control Systems* by J. DiStefano et al. (2011).

CHAPTER 26

Topics Beyond This Book

The last few chapters have offered a fairly comprehensivse sketch of what could be called basic or elementary feedback theory. Of course, there is much more that could be said.

Discrete-Time Modeling and the *z*-Transform

The theory presented here assumes that time is a *continuous* variable. This is not true for digital systems, where time progresses in discrete steps. When applying the continuous-time theory to such processes, care must be taken that the step size is smaller (by at least a factor of 5–10) than the shortest time scale describing the dynamics of the system. If this condition is not satisfied, then the continuous-time theory can no longer be safely regarded as a good description of the discrete-time system.

There is an alternate version of the theory that is based directly on a discrete-time model and that is generally useful if one desires to treat discrete time evolution explicitly. In discrete time, system dynamics are expressed as *difference equations* (instead of differential equations) and one employs the *z-transform* (instead of the Laplace transform) to make the transition to the frequency domain.

Structurally, the resulting theory is very similar to the continuous-time version. One still calculates transfer functions and examines their poles and zeros, but of course many of the details are different. For instance, for a system to be stable, all of its poles must now lie inside the *unit circle* around the origin in the (complex) *z*-plane, rather than on the left-hand side of the plane. And the entries in a table of transform pairs are different from those in Table 20-1, of course.

This discussion assumes that one is actually in possession of a good analytical model of the controlled system and intends to use the theory for calculating quantitative results! Rembember that (with the exception of the continuous-time cooling fan speed example) for *none* of the case studies in Part III did we have an analytical model of the system's time evolution at all, and theoretical results were meaningful only "by analogy." However, in cases where the controlled system exhibits nontrivial dynamics *and* there is a reasonably good analytical model, z-transform methods should be used if the sampling interval is not significantly shorter than the shortest relevant time scale of the controlled system.

State-Space Methods

All the theoretical methods discussed in preceding chapters were based on the transformation from the time domain into the frequency domain. In addition to these "classical" frequency-domain methods, there exists a completely different set of mathematical methods for the design of feedback control systems that is known as *time-domain, state-space*, or simply "modern" control theory. (These methods were developed in the 1960s.)

Whereas classical control theory relies on transforms to the frequency domain, state-space methods are based on linear algebra.[1] The theoretical development begins with the realization that any linear differential equation, regardless of its order, can be written as a *linear system of first-order equations* by introducing additional variables. As an example, consider the familiar second-order equation that describes a harmonic oscillator:

$$\frac{d^2}{dt^2} y(t) + 2\zeta\omega \frac{d}{dt} y(t) + \omega^2 y(t) = u(t)$$

If we introduce the new variable

---

1. This section presumes that the reader is familiar with linear algebra.

$$z = \frac{dy}{dt}$$

then the original second-order equation can be written as a system of two coupled first-order equations:

$$\frac{d}{dt} y = z$$

$$\frac{d}{dt} z = -2\zeta \omega z - \omega^2 y + u$$

Clearly, the process can be extended to equations of higher order if we introduce more variables. It is easy to treat vector-valued equations or systems of equations this way by turning each component of the original vector into a separate first-order equation.

These equations can be written in matrix form (where a dot over a symbol indicates the time derivative) as follows:

$$\begin{pmatrix} \dot{y} \\ \dot{z} \end{pmatrix} = \begin{pmatrix} 0 & 1 \\ -\omega^2 & -2\zeta\omega \end{pmatrix} \begin{pmatrix} y \\ z \end{pmatrix} + \begin{pmatrix} 0 \\ 1 \end{pmatrix} u$$

In general, any system of linear, first-order equations can be written in matrix form as

$$\dot{x} = Ax + Bu$$

where x and u are *vectors* and where A and B are matrices. Because there are always as many equations as variables, the matrix A is square; however, depending on the dimension of input u, the matrix B may be rectangular.

Such a system of linear first-order differential equations always has a solution in terms of the *matrix exponential*:

$$x(t) = e^{At} x_0$$

Here x_0 is a vector specifying the initial conditions, and the matrix exponential is defined via its Taylor expansion:

$$e^{At} = 1 + At + \frac{1}{2}A^2t^2 + \frac{1}{3!}A^3t^3 + \cdots$$

Now consider the following feedback system:

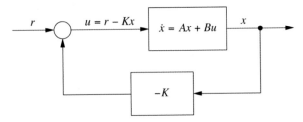

In contrast to what we have seen in the preceding chapters, the information about the system's dynamic behavior is *not* provided through its transfer function in the frequency domain. Instead, that information is specified explicitly through the system of differential equations.

The controller K is a matrix, yet to be determined, that acts on (is multiplied by) the system's output x. Given the setpoint r, we can therefore express the system's control input u as follows:

$$u = -Kx + r$$

Plugging this expression into the differential equation $\dot{x} = Ax + Bu$ that describes the system yields

$$\begin{aligned} \dot{x} &= Ax + B(-Kx + r) \\ &= (A - BK)x + Br \end{aligned}$$

This equation is the equivalent of the "feedback equation" (Chapter 21) for the state-space representation.

The matrix $(A - BK)$ determines the dynamics of the *closed-loop* system; specifically, the *eigenvalues* ω_j of this matrix correspond to the individual *modes* of the dynamical system (as discussed in Chapter 23). As mentioned earlier, the solution to the differential equation is given by the matrix exponential. Inserting the matrix (in diagonalized form) into the matrix exponential leads to terms of the form $e^{\omega_j t}$, so that ultimately the time evolution of the dynamical system is again represented as a linear superposition of harmonic terms.

So in order to design a closed-loop system that has the desired behavior, we must "assign the system's eigenvalues" to the appropriate values. (This is equivalent to "placing the system's poles" when working in the frequency domain.) Recall that A and B are determined by the system itself but that the controller K is still completely undetermined. It turns out that under certain conditions it is possible to move the eigenvalues of the closed-loop transfer matrix $(A - BK)$ to arbitrary locations by adjusting the entries of K. Moreover, it is possible to write down explicit expressions that yield the entries of K directly in terms of the desired eigenvalues.

This is a considerable achievement. By following the program just outlined, it is possible to design a closed-loop system having *any* desired dynamic behavior. Once the desired behavior has been specified (in terms of the eigenvalue positions), the controller can be calculated in a completely deterministic way.

It is instructive to compare this approach with frequency-domain methods. There, the *form* of the controller was fixed to be of the PID-type; the only means to adapt it to the particular situation was to find the best values for its two (or three) gain parameters. Now, the form of the controller is much less constrained (it is only required to be linear), and its entries are determined completely in terms of the desired behavior. Whereas frequency methods required the "tuning" of a predetermined controller, state-space methods amount to "designing" or "synthesizing" the controller from scratch.

Beyond their immediate application to controller design, state-space methods also allow for new ways to reason about control systems. In particular for systems involving multiple input and output signals, state-space methods enable new insights by bringing methods from linear algebra and matrix analysis to bear on the problem.

For systems involving multiple input and output channels, the question arises under what conditions the inputs are sufficient to establish control over all the outputs simultaneously. It is intuitively plausible that, if a system has fewer inputs than outputs, then in general it won't be possible to control the values of *all* the outputs simultaneously. State-space methods help to make this intuition more precise by relating it to questions about the rank of certain matrices that are derived from the matrices A and B determining the controlled system.

To summarize, state-space methods have several advantages over the classical theory:

- They facilitate total controller *synthesis* (as opposed to mere tuning).
- They are easily extendable to systems involving multiple input and output channels.
- They enable additional analytical insights based on operations from linear algebra.

At the same time, however, state-space methods also have some serious drawbacks:

- They *require* a good process model in the form of linear differential equations.
- They are quite abstract and easily lead to purely formal manipulations with little intuitive insight.
- Their results may not be *robust*.

The last point brings up the issue of *robustness*, which is the topic of the next section.

Robust Control

The controller-design program outlined in the previous section seemed foolproof: once the desired behavior has been specified in terms of a set of eigenvalue positions, the required controller is completely determined by a set of explicit, algebraic equations. What could possibly go wrong?

Two things, in fact. First of all, it is not clear whether a controller "designed" in this way is feasible from a technical point of view: the control actions required may be larger than what can be built with realistic equipment. But beyond these technical issues, there is a deeper theoretical problem: the controller that was found as solution to the algebraic design problem may be suitable only for *precisely* that particular problem as expressed in the differential equations describing the dynamical system. In reality, there will always be a certain amount of "model uncertainty." The physical laws governing the system may not be known exactly or some behavior was not included in the equations used to set up the calculation; in any case, the parameters used to "fit" the model to the actual apparatus are known only to finite precision. State-space methods can lead to controllers that work well for the system as specified yet perform poorly for another system—even if it differs only minimally from the original one.

The methods known as *robust control* address this issue by providing means to quantify the differences between different systems. They then extend the original controller design program to yield controllers that work well for all systems that are within a certain "distance" of the original system and to provide guarantees concerning the maximum deviation from the desired behavior.

This topic can quickly become quite involved.

Optimal Control

In addition to the momentary performance requirement (in terms of minimum rise time, maximum overshoot, and so on), it may be desirable to optimize the overall *cost* of operating the control system, typically over an extended period of time. We encountered this consideration in several of the case studies in Part III: if there is a fixed cost with each control action, then naturally we will want to reduce the number of distinct control actions. At the same time, there may also be a cost associated with the existence of a persistent tracking error, so that there is a need to balance these two opposing factors. This is the purview of *optimal control.*

In some ways, optimal control is yet another controller design or tuning method, whereby instead of (or in addition to) the usual performance requirements the system is also expected to extremize an arbitrary *performance index* or *cost function.* The performance index is typically a function that is calculated over an extended time period. For instance, we may want to minimize, over a certain time interval, the average tracking error or the number of control actions.

Optimal control leads naturally to the solution of optimization problems, usually in the presence of constraints (such as limits on the magnitude of the control actions, and so forth). These are difficult problems that typically require specialized methods. Finally, it is obviously critical to ensure that the performance index chosen is indeed a good representation of the cost to be minimized and that all relevant constraints are taken into account.

Mathematical Control Theory

As the last few sections suggest, the mathematical methods used to study feedback and control systems can become rather involved and sophisticated. The classical (frequency-space) theory uses methods from complex function theory to prove the existence of certain limits on the achievable performance of feedback systems. For example, the "Bode integral formula" states that systems cannot exhibit ideal behavior over the entire frequency range: an improvement in behavior for some frequencies will lead to worse behavior at other frequencies.

The endpoint of this line of study is *mathematical control theory*, which regards control systems as purely mathematical constructs and tries to establish their properties in a mathematically rigorous way. The starting point is usually the dynamical system $\dot{x} = Ax + Bu$ (or, more generally, $\dot{y} = f(y, u)$, and strict conditions are established under which, for example, all solutions of this system are bounded and thus indicate stability. The direct application of these results to engineering installations is not the primary concern; instead, one tries to understand the properties of a mathematical construct in its own right.

PART V
Appendices

Glossary

Actuator
> A device to convert a control signal (as produced by the controller) into a physical action that directly affects the plant. A heating element is an actuator, as is a stepper motor. Actuators are transducers.

Actuator saturation
> Because they are physical devices, actuators have limits in the action they can bring about. (A heating element has a maximum amount of heat it can generate per second, a motor has a maximum velocity, and so on. Notice in particular that a heating element is completely unable to generate any cooling action or negative heat flow.) At the same time, control signals can be arbitrarily large. Whenever an actuator is unable to follow the demands of the control signal, it is "saturated." Actuator saturation means that the intended control actions are no longer applied and that the control loop is therefore broken. (See also Integrator clamping.)

Bang-bang controller
> A colloquial term for an on/off controller (as opposed to a controller that is capable of varying the magnitude in response to its input).

Bumpless transfer
> A smooth transition when switching between different controllers —for instance between manual control and closed-loop control or between different control strategies in a gain-scheduling scenario. When using a PID controller, a bumpless transfer requires

that the value of the integral terms be synchronized before the transfer. (See also Integral preloading.)

Control problem
Given a system with an input and output, the control problem for this system amounts to finding the input setting (or sequence of input settings) that will produce a desired output value (or sequence of output values).

DC gain
Also known as "zero-frequency gain." This is the ratio of an element's output to a constant (zero-frequency) input while in the steady state (that is, after all transient behavior has disappeared).

Delay
Also known as "dead time." This is the time interval during which no response to an input change is visible in a system's output. (See also Lag.)

Derivative control
A controller whose output is proportional to the derivative of its input.

Derivative kick
When using a derivative controller, a sudden setpoint change will lead to a response from the derivative term that can—in principle—be infinitely large. This "kick" is usually not desirable.

Digital control
Any control strategy or implementation that uses digital controllers (as opposed to analog controllers made from physical devices). In a narrower sense, this term refers to control loops operating in discrete time steps.

Distributed parameter model
A model of a plant or process that requires an infinite set of parameters to describe the momentary state of the plant. (See also Lumped parameter model.)

Disturbance
Influences to the controlled system that cannot be controlled directly. (See also Load disturbance, Measurement noise.)

Disturbance rejection
The ability of the controller to maintain the output at the desired value, even in the presence of disturbances.

Dynamics

The dynamics of a system consist of the system's time evolution and its response to inputs.

Error feedback

A closed-loop control strategy in which the tracking error $e = r - y$ is used as input signal to the controller. Error feedback is subject to large control actions when the setpoint undergoes sudden changes. (See also Output feedback.)

Error-squared control

A controller in which the output is proportional to the square of its input (usually the error). The square must be calculated as $e \cdot |e|$ in order to retain information about the sign of the error. (See also Linear controller.)

Feedback control

Also known as "closed-loop control." This is a strategy for solving a control problem that is based on continuously comparing the actual process output against the reference value and then applying corrective actions to the input, in order to reduce the difference between the actual and the desired output. Because the actual process output is used in determining the new control input, feedback control "closes the loop" or introduces a "feedback path." (See also Feedforward control.)

Feedforward control

Any control strategy that does not take the actual process output into account when determining the new control input. Feedforward control requires relatively detailed knowledge about the behavior of the controlled system and cannot guarantee robustness to random, unforeseen disturbances. (See also Feedback control.)

Frequency domain

The time evolution of a system is described by functions of time t. The information contained in these functions can also be expressed by functions of complex frequency s. Switching between both representations is accomplished through transformations such as the Laplace transform. When considering only functions of complex frequency, one is working in the frequency domain. (See also Time domain.)

Gain scheduling

Sometimes different circumstances require different controller gains (for example, there may be a "heavy traffic" and a "light traffic" regime). A gain schedule is a table that contains appropriate values for the gain factors applicable to each separate regime.

Incremental controller
Also known as "velocity algorithm." This is a controller that calculates only the *change* in control signal. Incremental controllers can be used with plants that maintain their own state and respond to updates of their control input.

Input
Also known as "control input" or "manipulated variable." This is a quantity that influences the behavior of the controlled system and that can be manipulated directly.

Integral control
A controller whose output is proportional to the time integral of its past inputs.

Integrator clamping
Also known as "conditional integration." Integrator clamping means that the integral term inside a PI or PID controller is not updated when the actuator is saturated. In the case of actuator saturation, tracking errors may persist for a long time (since the actuator is unable to apply the control action required to eliminate the error and the system is therefore running in an open-loop configuration). Integrator clamping prevents the controller's integral term from becoming very large under these irregular (open-loop) conditions. (See also Integrator windup.)

Integrator preloading
The process of initializing or otherwise adjusting the cumulative or integral term in a PID controller outside of regular, closed-loop operations. (See also Bumpless transfer.)

Integrator windup
When the actuator has saturated, tracking errors may persist for a long time because the actuator is then unable to apply the control action required to eliminate the error. These persistent tracking errors will be added to the integral term inside the controller (unless the integrator is "clamped"). When the actuator is no longer saturated and the system is therefore operating in a closed-loop configuration again, the value of the integral term will nevertheless persist until it has been "unwound," resulting in inappropriate control actions. (Compare: Integrator clamping.)

Internal model controller
A controller that contains a model of the process (so that the model is "internal" to the controller) and uses the output from this model when computing control actions.

Lag
A system that initially responds with a partial response to an input change shows a lag. The term also refers to the duration until the system's output does replicate the input. (See also Delay.)

Level control
Control scenarios in which the system output is to be kept within a range of values rather than tracking a given reference value (setpoint) exactly. (Example: the fluid level in a storage tank is allowed to fluctuate as long as the tank neither overflows nor runs empty.)

Linear controller
A controller in which the output is a linear function of its input (which is typically the error): doubling the input results in a doubling of the output. The PID controller is a linear controller.

Load disturbance
A disturbance, such as noise, that affects the controlled system (that is, the plant or process). Since the controller provides the input to the plant, the plant constitutes the "load" that the controller must drive.

Loop shaping
The introduction of additional elements (such as filters or compensators) into a control loop with the intent of changing the loop's dynamic response.

Loop transfer function
Also known as "open-loop transfer function." If several elements (such as a controller and a plant) are arranged in a closed loop, then the transfer function of the corresponding *open loop* may be simply referred to as the loop transfer function.

Lumped parameter model
A model of a plant or process that describes the entire momentary state of the plant in a finite set of parameters. (See also Distributed parameter model.)

Manipulated variable
See Input.

Measurement noise
Noise generated in the sensor that is used to monitor the plant's output signal.

Model
A mathematical description (typically in the form of a differential equation) of the behavior of a plant or system, including the plant's dynamic behavior and its response to control inputs. (Before the advent of digital controllers, a "model" was a *physical model* of the plant, built to reproduce the actual plant's response to inputs.)

Model identification
See System identification.

Model reduction
The procedure by which a complicated mathematical model is replaced by a simpler one that nevertheless describes the observed behavior nearly as well. Basing a model on the theoretical knowledge about often results in models that are overly detailed. If experimental observations suggest that a simpler model will do as well, then one may attempt to simplify matters through model reduction.

Model uncertainty
As mathematical idealizations of a real system, models rarely describe the system perfectly. That imperfection introduces a certain amount of error into results, which is referred to as "model uncertainty." To an outside observer, it is impossible to tell whether the errors are due to inaccuracies in the description of the plant (model uncertainty) or to changing environmental factors (load disturbances).

Non-minimum phase system
Also known as "inverse response system." This is a system whose initial, transient response to an input is in the opposite direction to the input.

Open-loop transfer function
See Loop transfer function.

Output
Also known as "process output" or "process variable." This is the property of the controlled system that is to be influenced. Because it cannot be manipulated directly, the only way to influence it is by manipulating the controlled system instead.

Output feedback

A closed-loop control strategy in which the output y is used as input signal to the controller instead of the full tracking error $e = r - y$. Output feedback is less susceptible to spurious control actions in the case of sudden setpoint changes, but it gives equivalent results—in particular when the setpoint is held constant for extended periods of time. (See also Error feedback and Setpoint weighting.)

PI controller

Also known as "two-term controller." This is a linear controller that consists of a proportional and an integral term acting in parallel. The relative strength of each term is given by that term's controller gain factor.

PID controller

Also known as "three-term controller." This is a linear controller that consists of a proportional, an integral, and a derivative term acting in parallel. The relative strength of each term is given by that term's controller gain factor. The term "PID controller" is used even when one of the terms is absent (in particular, the derivative term is often not used).

Plant

Also known as "process." This is the system that needs to be controlled. The system's input is manipulated in order to achieve a particular behavior of the system's output.

Plant signature

See Process reaction curve.

Pole

A location at which a transfer function approaches infinity. Since transfer functions are typically rational functions in the complex frequency s, a pole is a value of s such that the denominator of the transfer function vanishes. A pole s in the frequency domain corresponds to a mode $\exp(st)$ in the time domain. Knowledge about the system's poles therefore amounts to knowledge about the system's modes in the time domain.

Pole placement

A process of manipulating the loop's transfer function in order to move its poles into positions that yield desirable dynamic behavior.

Process
See Plant.

Process characteristic
Also known as "static process characteristic." This is a curve showing the steady-state output of a process as function of the magnitude of the (constant) input.

Process control
The application of methods from control theory and engineering
to processes and installations in the chemical and manufacturing
industry.

Process knowledge
Knowledge about the static (steady-state) input/output relationship for a plant or process and about its dynamic response to arbitrary inputs. Process knowledge can be gained either analytically (if the laws governing the process are known) or empirically.
Process knowledge is captured in the process model. (See also
Process model, Model uncertainty, and System identification.)

Process model
A theoretical description of a plant or process—in particular of
its dynamic response to control inputs. Process models can describe a specific installation in detail; however, the term is also
used in an abstract sense to refer to broad categories of behaviors
(such as "self-regulating process," "accumulating process," "oscillatory process," and so on). (See also Model uncertainty.)

Process reaction curve
Also known as "plant signature." This is a curve showing the dynamic development of a process's output in response to a step
input.

Process variable
See Output.

Proportional control
A controller whose output is proportional to its input.

Proportional droop
Under strictly proportional control, the system's steady-state output will always be smaller than the setpoint. The higher the controller gain, the smaller the droop. In general, it is necessary to
employ integral control to entirely eliminate a steady-state tracking error.

Rate feedback
    See Velocity feedback.

Regulator
    Regulators are controllers that seek to maintain the system at its
    steady state and to reject disturbances. Regulators are used in sit-
    uations where the setpoint is constant for extended periods at a
    time. (See also Tracker.)

Sensor
    A device that transforms a physical quantity into a control signal.
    A thermocouple is a sensor that transforms temperature into
    voltage.

Servo-mechanism
    Also known as a "servo." See Tracker.

Setpoint
    Also known as "reference," "reference value," or "target." This is
    the desired value that the output of the controlled system is sup-
    posed to replicate.

Setpoint following
    The ability of a control system to track the setpoint accurately—
    especially in situations where the setpoint itself is undergoing
    changes.

Setpoint response
    Dynamic response of a controlled, closed-loop system to changes
    in the setpoint—in particular to sudden, steplike changes.

Setpoint weighting
    When calculating the tracking error $e = \alpha r - y$ that is to be used
    as controller input, the weight α of the setpoint r can be changed
    relative to the process output y. Choosing $\alpha = 1$ amounts to error
    feedback; choosing $\alpha = 0$ amounts to output feedback. (See also
    Error feedback and Output feedback.)

Smith predictor
    A control strategy to handle systems whose dynamic behavior
    exhibits a significant delay.

Steady state
    The behavior of the system after the disappearance of all transient
    responses. The steady-state output is usually dominated by the
    control inputs of the system. (See also Transient response.)

Step input
> An input that undergoes a sudden change in magnitude at a specific point in time (usually taken to be the beginning of the observation period, $t = 0$).

System identification
> The process of measuring a plant's behavior for the purpose of identifying and fitting a mathematical model.

Time domain
> The time evolution of a system is described by functions of time t. When considering the actual dynamic behavior of a system, one is working in the time domain. (See also Frequency domain.)

Tracker
> A controller designed to follow a setpoint that is changing over time. (See also Regulator.)

Transducer
> Any device that converts between physical actions or quantities and control signals. Actuators and sensors are transducers. Control signals are often, but not always, electrical signals.

Transfer function
> The frequency-space representation of the laws governing an element's dynamics. Transfer functions are usually obtained through the Laplace transformation of the differential equation describing the element. To obtain the response of the element to an arbitrary input, the transfer function is multiplied by the Laplace transform of the input signal; the resulting product is then transformed back into the time domain to obtain the dynamic response.

Transient response
> Also known as "transients" or "transient behavior". This is that part of the dynamic response to an input change that decays and disappears over time. Transients are usually due to the internal dynamics of the controlled system, not to its control inputs. (See also Steady state.)

Velocity algorithm
> See Incremental controller.

Velocity feedback

    Also known as "rate feedback." This is a closed-loop control strategy in which the rate of change of the plant's output (that is, its derivative) is fed back and used to calculate the new control input. (Not to be confused with "Velocity algorithm.")

Creating Graphs with Gnuplot

Gnuplot (*www.gnuplot.info*) is an open source program for plotting data and functions. It is intended primarily for Unix/Linux systems, although versions for Windows and the Mac exist as well.

Basic Plotting

When started, gnuplot provides a shell-like interactive command prompt. All plotting is done using the `plot` command, which has a simple syntax. To plot a function, you would type (at the gnuplot prompt):

```
plot sin(x)
```

By default, gnuplot assumes that data is stored in white-space separated text files, with one line per data point. To plot data from a file (residing in the current directory and named `data`), you would use the following:

```
plot "data" using 1:2 with lines
```

This assumes that the values used for the horizontal position (the x values) are in the first column, and the values for the vertical position (the y values) are in the second column. Of course, any column can be used for either x or y values.

Gnuplot makes it easy to combine multiple curves on one plot. Assume that the data file looks like the one shown in Figure B-1 (left). Then the following command would plot the values of both the second and the third column, together with a function, in a single graph (Figure B-1, right):

```
plot "data" u 1:2 w lp, "data" u 1:3 w lp, 5*exp(-5*x)
```
Here we have made use of the fact that many directives can be abbreviated (u for using and w for with) and have also introduced a new plotting style, linespoints (abbreviated lp), which plots values as symbols connected by lines. The other two most important styles are lines, or l, which connects data points with straight lines but does not plot symbols, and points, or p, which plots only symbols but no lines.

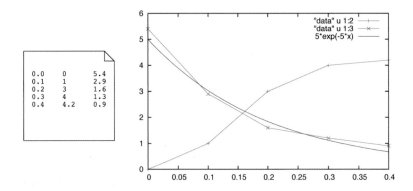

Figure B-1. A data file (left) and the resulting plot (right).

Plot Ranges

It is possible to modify the ranges of values included in a plot ("zooming") by using range specifiers as follows.

```
plot [0:0.5] plot "data" using 1:2 w p
plot [0:0.5][0:7] plot "data" using 1:2 w lp
plot [][0:7] plot "data" using 1:2 w l
```

The first line would limit the *x* range, the second line would limit both the *x* range and the *y* range, and the third line would limit *only* the *y* range.

Inline Transformations

Gnuplot is, by design, only a *plotting* utility; it is not a general-purpose computational workbench (like Matlab, Octave, R, and many others). Nevertheless, it is possible and often useful to apply a transformation to data points as they are being plotted.

The following code will plot the square root of the second column:

```
plot "data" using 1:(sqrt($2)) with lines
```

Whenever the column specification in the using directive is enclosed in parentheses, the contents of the parentheses are evaluated as a mathematical expression. When inside parentheses, you can access column values by preceding the column number with a dollar sign ($). The basic arithmetic operations are supported, as are all standard (and some not-so-standard) mathematical functions.

Plotting Simulation Results

The simulation framework introduced in Chapter 12 writes results out (to standard output) in a format that is suitable for gnuplot. Assuming that standard output has been redirected to a file called results, the setpoint (column 3) and the actual plant output (column 7) can be plotted as a function of the wall-clock time (column 2) with a command like this:

```
plot "results" using 2:3 with lines, "" u 2:7 w l
```

Notice how the filename has been left blank the second time, in which case gnuplot reuses the most recently given filename.

Summary

These instructions should be sufficient to allow you to make basic plots. Extensive documentation is available within a gnuplot session using the help command:

```
help plot
```

The same information can also be found as a PDF on the gnuplot website.

In addition to the basic plot command, gnuplot provides myriad of *options* to change the appearance of a plot or to add annotations to it. A systematic introduction can be found, for instance, in my book on gnuplot.[1]

---

1. *Gnuplot in Action: Understanding Data with Graphs* by Philipp K. Janert (2009)

Complex Numbers

Each complex number z is a point in the complex plane, which is spanned by the real axis and the imaginary axes:

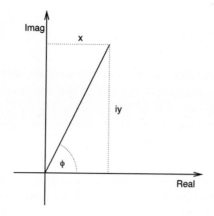

Two coordinate systems are commonly used for a (two-dimensional) plane: Cartesian and polar coordinates. For every complex number there exist two equivalent representations:

$$z = x + iy \qquad \text{Cartesian}$$
$$z = r\, e^{i\phi} \qquad \text{Polar}$$

Here

$$i = \sqrt{-1} \quad \Longleftrightarrow \quad i^2 = -1$$

is the "imaginary unit."

We can transform between those representations as follows:

$$x = r \cos(\phi) \qquad \text{Real part}$$

$$y = r \sin(\phi) \qquad \text{Imaginary part}$$

$$r = \sqrt{x^2 + y^2} \qquad \text{Magnitude}$$

$$\phi = \arctan\left(\frac{y}{x}\right) \qquad \text{Phase}$$

Basic Operations

Complex numbers are added and multiplied component by component while taking into account that $i^2 = -1$. If $z_1 = x_1 + iy_1$ and $z_2 = x_2 + iy_2$, then

$$z_1 + z_2 = (x_1 + x_2) + i(y_1 + y_2)$$

$$z_1 z_2 = (x_1 x_2 - y_1 y_2) + i(x_1 y_2 + x_2 y_1)$$

Each complex number z has a "complex conjugate," denoted $z^\star$, which is the same as z except that the sign of the imaginary part has been reversed. Thus,

$$z = x + iy \qquad \Longleftrightarrow \qquad z^\star = x - iy$$

The product of a complex number and its conjugate is always real. The square root of this product is also real. It is called the magnitude of z and is denoted by $|z|$.

$$zz^\star = (x + iy)(x - iy) \quad = \quad x^2 + y^2$$

$$|z| = \sqrt{zz^\star} \qquad\qquad = \quad \sqrt{x^2 + y^2}$$

The following identities hold:

$$|z_1 z_2| = |z_1| \cdot |z_2| \quad \text{and} \quad \left|\frac{z_1}{z_2}\right| = \frac{|z_1|}{|z_2|}$$

We can use the complex conjugate to assign meaning to the reciprocal of a complex number:

$$\frac{1}{z} = \frac{z^\star}{zz^\star} = \frac{z^\star}{|z|^2}$$

Polar Coordinates

Every complex number z can also be expressed in polar coordinates:

$$z = r\,e^{i\phi}$$

where

$$r = |z|$$

$$\phi = \arg z = \arctan\frac{\text{imag}}{\text{real}}$$

The radius r can be included in the argument of the exponential:

$$r\,e^{i\phi} = e^{\log(r)}e^{i\phi} = e^{\sigma+i\phi} \qquad \text{where} \qquad \sigma = \log(r)$$

In polar coordinates, multiplication of two complex numbers amounts to multiplying the magnitudes and adding the phases:

$$z_1 \cdot z_2 = r_1 e^{i\phi_1} \cdot r_2 e^{i\phi_2} = r_1 r_2 e^{i(\phi_1+\phi_2)}$$

Taking the complex conjugate of a complex number is equivalent to changing the sign of its phase:

$$z = r e^{i\phi} \qquad \Longleftrightarrow \qquad z^\star = r e^{-i\phi}$$

When multiplying z and $z^\star$, the exponential terms cancel and so leave the purely real number r^2.

The Complex Exponential

If a complex number z is expressed in polar coordinates,

$$z = r\,e^{i\phi}$$

then all information about the *magnitude* of the number is contained in the radius r. The exponential term $e^{i\phi}$ provides information about the number's angular *orientation* as a point in the complex plane. For this reason, the exponential term is also known as the *phase factor*.

A phase factor is an exponential term with a purely imaginary exponent. Its magnitude is always 1:

$$|e^{i\phi}| = 1$$

Geometrically, the phase factor describes a point on the unit circle. A straightforward geometric construction allows us to express it in terms of trigonometric functions:

$$e^{i\phi} = \cos(\phi) + i\sin(\phi)$$

For multiples of $\pi/2 = 90$ degrees, the phase factor takes on special values:

$$e^{i\cdot0} = 1 \qquad e^{i\pi/2} = i \qquad e^{i\pi} = -1 \qquad e^{i3\pi/2} = -i$$

Of particular importance is the case when the phase angle grows steadily with time: $\phi = \omega t$, where ω is a real constant (the "angular frequency").

In this case, the phase factor describes a purely oscillatory behavior at constant amplitude. The two trigonometric functions simply "wiggle" and exhibit no growth or decay. The greater is ω, the faster are the wiggles:

$$e^{i\omega t} = \cos(\omega t) + i\sin(\omega t)$$

Further Reading

It can be difficult and frustrating for the practitioner to find background information about control theory that is relevant to one's problems. Most textbooks on control theory are intended mostly to prepare students to perform calculations using Laplace transform methods (or state-space techniques), while conceptual development and practical matters are given rather short shrift. On the other hand, practice-oriented titles are often little more than field guides that offer only heuristic rules of thumb for specific, well-known applications. Books or articles concentrating on feedback *concepts* are rare.

Recommended Reading

- *The Art of Control Engineering* by Ken Dutton, Steve Thompson, and Bill Barraclough (1997). This comprehensive volume provides an excellent introduction to control theory that balances theory and practical aspects. Be aware that the arrangement of topics within the book can make it difficult to follow the conceptual development. (If you have difficulties finding this book in the U.S., then look for it in the UK.)

- *Advanced PID Control* by Karl Johan Åström and Tore Hägglund (2005). Despite its seemingly narrow title, this is an extremely comprehensive and accessible book on the practical problems that one is likely to encounter when developing real-world feedback systems. Although the text is very "hands on" and geared toward field work (mostly in the chemical industry), it provides much more than just heuristic rules of thumb. This book is not an introduction, but it should possibly be everyone's *second* book on feedback systems. (The previous edition, entitled *PID Controllers: Theory, Design, and Tuning*, remains in print and is in some ways a more practical book.)

Additional References

- *Essentials of Control* by J. Schwarzenbach (1996). This mercifully short volume (only 140 pages) provides an easy-to-read introduction to feedback control. It covers similar material as the present book, but from a more scholarly perspective. Lots of worked math examples. (If you have difficulties finding this book in the U.S., look for it in the UK.)

- *Feedback Systems: An Introduction for Scientists and Engineers* by Karl Johan Åström and Richard M. Murray (2008). This book is intended as a modern, comprehensive introduction to all things feedback, but the result is rather strange. The authors spend an inexplicable amount of space pursuing various preliminaries and obscure side topics even as fundamental material receives short shrift or is omitted altogether. An oversupply of disparate examples, and the attempt to treat frequency and state-space methods simultaneously, end up confusing the reader. There is a lot of valuable material here, but it can be difficult to pick out. (An unfortunate number of production glitches has slipped into the printed book, but a PDF of the complete text, including fixes to known errata, is freely available from the website of one of the authors (*http://www.cds.caltech.edu/~murray/amwiki/index.php*).)

- *Feedback for Physicists: A Tutorial Essay on Control* by John Bechhoefer (*Reviews of Modern Physics*, volume 77 (2005), pages 783–836). This is a review article for physicists, by a physicist. It takes a long view, emphasizing concepts rather than technical or mathematical detail. It is difficult reading but covers some topics not found elsewhere. Most valuable for its outsider's point of view.

- *Schaum's Outline of Feedback and Control Systems* by Joseph DiStefano, Allen Stubberud, and Ivan Williams (2nd edition, 2011). Not a bad, if terse, introduction, with many worked problems. Covers classical (frequency) methods only, but it does so in detail.

- *Modern Control Engineering* by Katsuhiko Ogata (5th edition, 2009). A standard, college-level textbook on control theory.

- *Control System Design: An Introduction to State-Space Methods* by Bernard Friedland (2005). A compact and affordable introduction to state-space methods.

- *Feedback Control of Computing Systems* by Joseph L. Hellerstein, Yixin Diao, Sujay Parekh, and Dawn M. Tilbury (2004). Applications of feedback methods to computer systems. Not for beginners.

- *Feedback Control Theory* by John C. Doyle, Bruce A. Francis, and Allen R. Tannenbaum (1990). This book is not an introduction to feedback methods. Rather, it is a mathematical research monograph in which the authors summarize some of their results. A PDF of the complete book is available at no cost. (*http://www.control.utoronto.ca/people/profs/francis/dft.pdf*)

Mathematical Prerequisites

- *Complex Variables and Applications* by James Ward Brown and Ruel V. Churchill (8th edition, 2008). There are many good books on complex analysis. This (relatively) short volume provides a concise yet accessible introduction. The treatment seems especially suitable for an application-minded audience.

- *Fourier Analysis and Its Applications* by Gerald B. Folland (2009). This book is an outstanding introduction to transform methods (Fourier and Laplace transforms) and their application to differential equations. The presentation is very accessible and hands-on. (This book does require solid knowledge of complex numbers.)

- *An Introduction to Probability Theory and Its Applications, Volume 1* by William Feller (3rd edition, 1968). A classic treatment of basic probability theory.

Index

A

accumulating process model, 86–88
actuator saturation, 101–103
actuators framework (simulation), 132–133
ad delivery simulation, 149–160
  control loops for, 151–153
  cumulative goals for, 156–157
  gain scheduling and, 157–158
  performance, improving, 153–155
  system dynamics, measuring, 150–151
  weekend effect on, 158
AMIGO method, 97–98, 179
architecture, 113–120
  cascade controls, 117–118
  feedforward and, 116
  for finite resources, 195–198
  gain scheduling, 114–116
  nested controls, 117–118
  Smith predictor, 118–120

B

block-diagram algebra, 74, 219–225
  composite systems and, 219–220
  feedback equations and, 221–224

block-diagrams, 219
  arrows, 40
  rules for, 41
Bode plots, 261–263
  alternative graphical techniques to, 265–266
  appearance of, 263
  examining, 261
boxes, in block diagrams, 40
buffers, in control simulation, 173–183
bumpless transfer, 103

C

cache hit rate simulator, 137–148
  application to physical word, 140–141
  components of, 137–140
  controller tuning, 143–145
  system characteristics, measuring, 141–143
cascade controls, 117–118
  queue controls as, 175
  setup for, 175–178
  tuning, 175–178
cascaded control loops, 175

We'd like to hear your suggestions for improving our indexes. Send email to index@oreilly.com.

E

eigenvalues, assigning, 271
enterprise systems, feedback systems and, 24
error-square controllers, 109
"Evans" Rules for root locus diagrams, 247–249
Evans, W. R., 247

F

feedback
    flow control and, 10–13
    performance, improving, 153–155
feedback control
    basic idea, 52
    goal of, 10
    situations, 10
feedback equations
    alternative derivations of, 223–224
    block-diagram algebra and, 221–224
feedback loops, theory of, 35
feedback package, 128, 133
feedback systems, 15–25
    change in, 22–24
    corrective actions, 18–22
    discrete-time modeling, 267–268
    enterprise systems and, 24
    feedforward systems vs., 23–24
    flow control and, 10–13
    mathematical control theory, 274
    optimal controllers, 273
    robust controllers, 272–273
    setpoint, 21–22
    signals and, 16–18
    state-space methods, 268–272
    tracking error, 18–22
    uncertainty, 22–24
    z-transform, 267–268
feedforward systems, 10
    control strategy, 116
    employing, 116
    large disturbances and, 116
    using, 116
filters, 132–133

finite resources, 193–201
    analyzing requirements for, 194–195
    architecture options for, 195–198
    flow, 3–13
    approaches to, 5–6
    establishing control of, 7–8
    feedback controls for, 10–13
    small deviations in, 8–9
framework (simulation), 129–136
    actuators/filters, 132–133
    components, 130–130
    controllers, 130–131
    graphical output, generating, 135–136
    plants/systems, 130
    standard loops, 133–135
frequency response, 257–266
    Bode plot and, 261–263
    for transfer functions, 258
    in physical world, 258
    stability criterion for, 263–265
frequency space, 210
functional outputs, controls vs., 57

G

gain controller, 91, 94
gain crossover frequency, 265
gain margin, 265
gain scheduling, 114–116
    applications, 114
    dynamic pricing and, 157–158
    nonlinear systems and, 115–116
gap nonlinear controllers, 109–110
global structure, 250
granularity, of control inputs, 64

H

harmonic oscillators, 217–218
hit rates, 32
    maintaining, 137

I

identifying properties, rules for, 266
identity elements, 132